Instability, nonexistence and weighted energy methods in fluid dynamics and related theories

B Straughan
University of Glasgow

Instability, nonexistence and weighted energy methods in fluid dynamics and related theories

Pitman Advanced Publishing Program
BOSTON · LONDON · MELBOURNE

PITMAN BOOKS LIMITED
128 Long Acre, London WC2E 9AN

PITMAN PUBLISHING INC
1020 Plain Street, Marshfield, Massachusetts 02050

Associated Companies
Pitman Publishing Pty Ltd, Melbourne
Pitman Publishing New Zealand Ltd, Wellington
Copp Clark Pitman, Toronto

First published 1982

AMS Subject Classifications: (main) 73BXX, 76D05, 35K60
(subsidiary) 35R25, 73B05, 73C50

Library of Congress Cataloging in Publication Data

Straughan, B. (Brian)
Instability, nonexistence and weighted energy methods in fluid dynamics and related theories.

(Research notes in mathematics; 74)
Bibliography: p.
1. Continuum mechanics. 2. Stability.
3. Differential equations, Partial. 4. Fluid dynamics. I. Title. II. Series. III. Weighted energy methods.
QA808.2.S77 1982 531 82-13168
ISBN 0-273-08564-6

British Library Cataloguing in Publication Data

Straughan, B.
Instability, nonexistence and weighted energy methods in fluid dynamics and related theories.
—(Research notes in mathematics; 74)
1. Differential equations, Partial—Solution
I. Title II. Series
515.3′53 QA374

ISBN 0-273-08564-6

ISBN 0 273 08564 6

Reproduced and printed by photolithography in Great Britain by Biddles Ltd, Guildford

Preface

This book presents an integrated account of modern developments in techniques for instability, nonexistence, stability and uniqueness in partial differential equations as applied to continuum mechanics, with particular emphasis on fluid dynamics. It began as two separate developments; one an attempt to review the literature on weighted energy methods, the other stemming from a course of lectures on nonexistence I gave in 1980 at the University of Strathclyde. There are many ideas common to both areas, however, and so it seemed sensible to combine them in this article.

I should like to thank Professor A.Jeffrey for his kind invitation to contribute a volume in the present series.

I am, naturally, deeply grateful to the friends and teachers to whom my understanding of the ideas described herein owes much: special thanks are due to G.P.Galdi, R.N.Hills, R.J.Knops, H.A.Levine, L.E.Payne and P.H.Roberts, F.R.S.

Throughout the book we use the IBM Courier symbol R to denote the real line.

Bearsden
May, 1982

B.Straughan

Contents

1 Introduction

In this research note we shall illustrate several techniques which lead to nonexistence results for solutions to partial differential equations. In addition, we shall describe several variations of a weighted energy method and show how these have been used to obtain results of stability, instability, uniqueness, nonexistence, asymptotic behaviour and a priori bounds for a variety of problems in mathematical physics. The list of references is not meant to be exhaustive; it does, however, cover fairly well the ideas we shall be interested in.

In connection with the opening sentence, one question immediately springs to mind, that is; what do we mean by nonexistence? To begin to answer this and to indicate what sort of things may happen we commence with two simple, but nonetheless enlightening, examples from ordinary differential equations.

a. Nonexistence by "blow up".

Consider the initial value problem

$$\begin{aligned} y' &= t + y^2, \quad t > 0, \\ y(0) &= 1, \end{aligned} \tag{1.1}$$

where $y' \equiv dy/dt$. First of all we observe that the right hand side of this equation is Lipschitz continuous in y and so there is at least local existence of the solution y. Now $y(0)=1$ and

$y' > 0$ for $t>0$ and so for continuously differentiable y, by continuity we conclude $y' > 0$, for each $t \geqq 0$. Also, since $t \geqq 0$, $y' \geqq y^2$ and so we may separate variables to find

$$t \leq \int_{y(0)}^{y(t)} \frac{dy}{y^2} \leq \int_1^{\infty} \frac{dy}{y^2} = 1, \tag{1.2}$$

since $y(0) \leqq y(t)$ and y^{-2} is positive. Therefore, we conclude $y(t)$ cannot exist beyond $t=1$. Furthermore,

$$y' \leq 1 + y^2, \qquad t \leq 1,$$

and so by a similar argument

$$\int_1^{y(t)} \frac{dy}{1+y^2} \leq t \ ,$$

i.e.

$$\tan^{-1} y(t) - \tan^{-1}(1) \leq t.$$

Since tan is positive and increasing on $[\pi/4, \pi/2)$, it follows that

$$y(t) \leq \tan(\pi/4 + t).$$

Hence we may conclude also that y must exist for t beyond $\pi/4$. Therefore, the behaviour of y is something like that shown in the diagram.

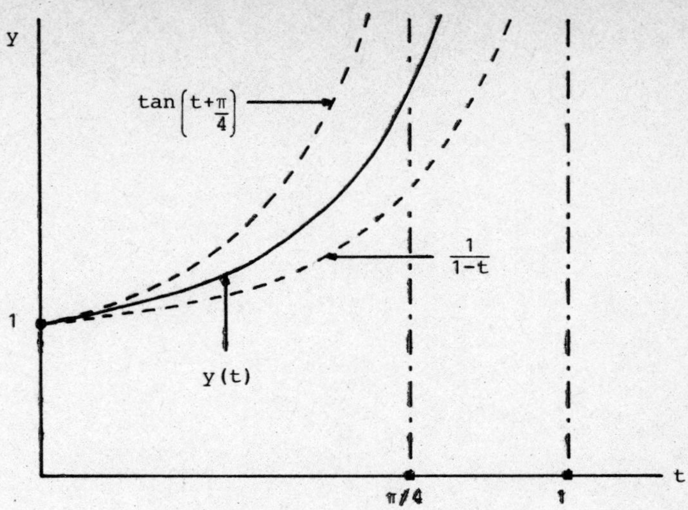

Figure 1.1

(It is easy to show from (1.2) that $y(t) \geqq 1/(1-t)$; however, the method illustrated generalizes more easily.)

We may actually conclude that $y(t)\to\infty$ in this example using the continuation properties of solutions to ordinary differential equations. However, when one is dealing with partial differential equations the situation is more delicate; while the same techniques allow us to derive nonexistence the establishment of "blow up" is more difficult (indeed, in some cases blow up does not occur, see Ball [3]).

As an illustration of another thing which may happen we give an example where the solution appears to decay to zero in a finite time.

b. Nonexistence by "decay to zero".

Dr. P. Calow of the Zoology department, University of Glasgow, has kindly shown me the following model he uses to predict the weight of a type of worm undergoing starvation.

$$\frac{dw}{dt} = aw + bw^{\alpha} + cw^{\beta}, \quad t > 0, \qquad (1.3)$$
$$w(0) = 1.$$

Here, $w(t)$ is the weight of the worm at time t, a, b, c, are negative constants and α, β are constants such that $0<\alpha<\beta<1$; typically, $\alpha=0.63$, $\beta=0.67$. It is convenient to take the initial weight w_o to be 1 as we may always redefine (1.3) with a new variable w/w_o and obtain a similar system to (1.3).

An explicit solution to (1.3) does not seem possible (although we can always integrate it implicitly numerically), but we can predict the behaviour of $w(t)$ by finding "upper and lower solutions" to (1.3). The idea is to introduce approximating solutions w_α, w_β where

$$\frac{dw_\alpha}{dt} = aw_\alpha + (b+c)w_\alpha^\alpha, \quad w_\alpha(0) = 1,$$
$$\frac{dw_\beta}{dt} = aw_\beta + (b+c)w_\beta^\beta, \quad w_\beta(0) = 1. \qquad (1.4)$$

Both of these equations are Bernoulli equations and so we may solve them (divide by w_μ^μ and obtain a linear equation for $w_\mu^{1-\mu}$, $\mu=\alpha$ or β) to find

$$w_\alpha(t) = [e^{a(1-\alpha)t} + \left(\frac{b+c}{a}\right)\{e^{a(1-\alpha)t}-1\}]^{1/(1-\alpha)},$$
$$w_\beta(t) = [e^{a(1-\beta)t} + \left(\frac{b+c}{a}\right)\{e^{a(1-\beta)t}-1\}]^{1/(1-\beta)}. \qquad (1.5)$$

Since $a,b,c<0$, it follows from (1.3) that $w(t) \leqq 1$, and so since $0<\alpha<\beta<1$,

$$w^\alpha \geqq w^\beta, \quad bw^\alpha \leqq bw^\beta, \quad cw^\alpha \leqq cw^\beta. \qquad (1.6)$$

Hence,

$$aw + (b+c)w^{\beta} \geqq aw + bw^{\alpha} + cw^{\beta} \quad (= \frac{dw}{dt}) \tag{1.7}$$
$$\geqq aw + (b+c)w^{\alpha},$$

and so if we solve the problems $dw/dt = aw+(b+c)w^{\beta}$ and $dw/dt = aw+(b+c)w^{\alpha}$ and denote their solutions by w_{β}, w_{α}, respectively, we have $w_{\alpha}(t) \leqq w(t) \leqq w_{\beta}(t)$, where w_{α}, w_{β} are given by (1.5). (This follows since from (1.7) $\frac{dw_{\beta}}{dt} \geqq \frac{dw}{dt} \geqq \frac{dw_{\alpha}}{dt}$ and $w_{\alpha}(0) = w(0) = w_{\beta}(0) = 1$.)

We know that $a,b,c<0$ and so from (1.5) $w_{\alpha}(t)$ and $w_{\beta}(t)$ decrease monotonically to zero in a finite time; in fact, with μ denoting either α or β,

$$w_{\mu} \to 0 \quad \text{as} \quad t \uparrow T_{\mu}, \quad \text{where} \tag{1.8}$$
$$T_{\mu} = \log\left(\frac{a+b+c}{b+c}\right)^{1/|a|(1-\mu)}.$$

Therefore, w(t) must either cease to exist before it reaches zero or become zero in a finite time, as shown below:

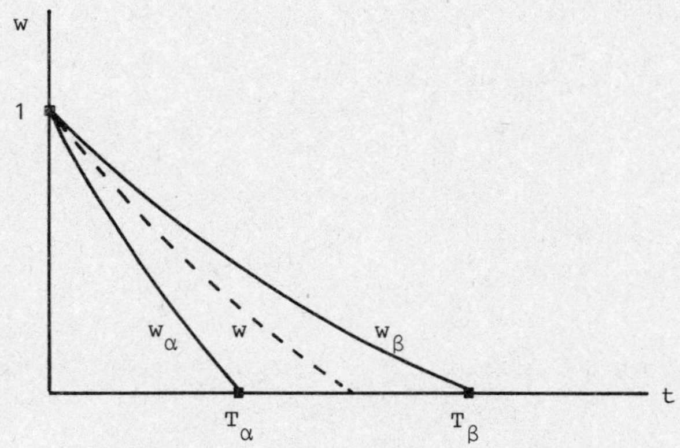

Figure 1.2

(Broken line indicates the possible behaviour of w.)

From a biological point of view, it probably means that the worm dies before $w=0$ is reached.

It is worth noting that we cannot assert w reaches 0 from the above argument; for the nonlinearity in (1.3) is not Lipschitz continuous and so we do not necessarily have local existence.

Other interesting nonexistence results for ordinary differential equations from boundary layer theory may be found in Bushell [11], Coppel [20], McLeod [124] and Reuter & Stewartson [161].

The above examples serve to illustrate what nonexistence behaviour we may expect in partial differential equations and in §§2-8 we describe several techniques which have been introduced to establish such nonexistence results.

To understand what is meant by a weighted energy method we first recall the conventional energy method in connection with uniqueness of the solution to the boundary-initial value problem for the heat equation.

Let Ω be a bounded domain in R^n with boundary Γ. Denote by $\hat{\Omega}$ the exterior of the closure of Ω, $\bar{\Omega}$. Suppose u is a solution to the boundary-initial value problem

$$\begin{aligned} u_t &= \Delta u, && \text{in } \Omega\times(0,\infty), \\ u &= 0, && \text{on } \Gamma\times[0,\infty), \\ u &\equiv 0, && \text{when } t = 0, \end{aligned} \tag{1.9}$$

where Δ is the n-dimensional Laplacian operator.

The standard energy method proceeds by deriving a differential inequality for the function E defined by

$$E(t) = \tfrac{1}{2}\int_{\Omega} u^2 dx.$$

To see this, multiply $(1.9)_1$ by u and integrate over Ω. We find

$$\frac{d}{dt}\left(\tfrac{1}{2}\int_{\Omega} u^2 dx\right) = \int_{\Gamma} n_j u u_{,j} dS - \int_{\Omega} u_{,i} u_{,i} dx, \qquad (1.10)$$

where $u_{,i} \equiv \partial u/\partial x_i$ and the Einstein summation convention is employed (eg. $u_{,i}u_{,i} = u_{,1}u_{,1}+u_{,2}u_{,2}+\ldots+u_{,n}u_{,n}$). Since the Dirichlet integral (i.e. $\int_{\Omega} u_{,i}u_{,i}dx$) is non-negative and u=0 on Γ, from (1.10) we deduce that

$$\frac{dE}{dt} \leqq 0,$$

and so

$$E(t) \leqq E(0) = 0.$$

E is a positive-definite functional of u and so necessarily $u \equiv 0$ in $\Omega\times[0,\infty)$.

We may now raise two questions:

1. What happens if in (1.9) the time interval $(0,\infty)$ is changed to $(-\infty,0)$? i.e. we consider the problem backward in time.
2. Does the energy method work when Ω is replaced by $\hat{\Omega}$?

To study question (1), note that (1.9) may be converted to

$$\begin{aligned} &u_t + \Delta u = 0, && \text{in } \Omega\times(0,\infty), \\ &u = 0, && \text{on } \Gamma\times[0,\infty), \\ &u \equiv 0, && \text{when } t = 0. \end{aligned} \qquad (1.11)$$

Instead of (1.10) we now obtain

$$\frac{dE}{dt} = \int_{\Omega} u_{,i}u_{,i}dx,$$

and so we are no longer able to deduce that $E(t) \leq 0$. Instead we must seek an alternative route. One such procedure is to consider a weighted energy,

$$\int_{\Omega} h(x,t)u^2 dx,$$

and then by suitable choice of h coupled with appropriate manipulation, deduce $u \equiv 0$ as before. For (1.11) this is not trivial and we defer the analysis until §11.

Consider now question (2). Then, the problem appropriate to (1.9) is

$$\begin{aligned} u_t &= \Delta u, && \text{in } \hat{\Omega}\times(0,\infty), \\ u &= 0, && \text{in } \Gamma\times[0,\infty), \qquad\qquad (1.12) \\ u &= 0, && \forall x \in \hat{\Omega}, \text{ when } t = 0. \end{aligned}$$

Suppose $0 \in \Omega$ and let $B(0,R)$ be the open ball centre 0, radius R. Choose R^* so that $\bar{\Omega} \subset B(0,R^*)$. Then define, for each $R \geq R^*$, $\Omega_R = B(0,R) - \bar{\Omega}$. Let $S(0,R)$ be the surface $x_i x_i = R^2$. If, as in the standard energy method, we multiply $(1.12)_1$ by u and integrate over Ω_R we find, analogous to (1.10)

$$\frac{d}{dt}\left(\tfrac{1}{2}\int_{\Omega_R} u^2 dx\right) = \int_{\Gamma \cup S(0,R)} u_{,i}un_i dS - \int_{\Omega_R} u_{,i}u_{,i}dx. \qquad (1.13)$$

The idea would be to define $E(t) = \frac{1}{2}\int_{\hat{\Omega}} u^2 dx$ and then let $R \to \infty$.

However, unless we impose decay conditions on the term

$$\int_{S(0,R)} n_i u u_{,i} dS$$

we are again unable to conclude $E(t) \leqq 0$. Thus, unless decay restrictions at infinity are imposed on u <u>and</u> $u_{,i}$ the conventional energy method again breaks down. (An additional decay condition on $u_{,i}$ is essential as may be seen by elementary examples such as in R if $u = \cos(x^m)/x$. Clearly $u = O(1/x)$ as $x \to \infty$, but du/dx may grow like x^{m-2} at infinity.)

Let us see how by using a weighted energy method one is able to overcome such problems. Let u be a solution to (1.12) and instead of decay conditions at infinity we impose the weak growth requirement that

$$|u|, \left|\frac{\partial u}{\partial r}\right| \leqq Me^{\lambda r}, \quad \text{as } r(=[x_i x_i]^{\frac{1}{2}}) \to \infty, \tag{1.14}$$

for $M(>0)$ constant and <u>any</u> preassigned $\lambda(>0)$.

Define a weighted energy E by

$$E(t) = \tfrac{1}{2}\int_{\hat{\Omega}} g u^2 dx,$$

where

$$g = e^{-\alpha r},$$

for $\alpha(>0)$ to be chosen. Then, multiply $(1.12)_1$ by gu and integrate over Ω_R. This leads to

$$\frac{d}{dt}\left(\tfrac{1}{2}\int_{\Omega_R} g u^2 dx\right) = \int_{\Omega_R} g u \Delta u dx,$$

$$= \int_{\Gamma\, S(0,R)} gun_i u_{,i} dS - \int_{\Omega_R} gu_{,i}u_{,i} dx - \int_{\Omega_R} g_{,i}uu_{,i} dx. \quad (1.15)$$

By calculation,

$$g_{,i} = -\alpha g \frac{x_i}{r} ,$$

and so we use the inequality

$$a_i b_i \leqq \frac{1}{2\omega} a_i a_i + \tfrac{1}{2}\omega b_i b_i , \quad \omega>0,$$

on the last term in (1.15) as follows

$$g_{,i}uu_{,i} \leqq \frac{\alpha^2}{4} gu^2 + gu_{,i}u_{,i} ,$$

from which results

$$\frac{d}{dt}\left(\tfrac{1}{2}\int_{\Omega_R} gu^2 dx\right) \leqq \int_{S(0,R)} gu\frac{\partial u}{\partial r} dS + \frac{\alpha^2}{4}\int_{\Omega_R} gu^2 dx. \quad (1.16)$$

Now, choose α so that

$$\alpha > 2\lambda, \quad (1.17)$$

where λ is the constant in (1.14). When we allow $R\to\infty$ in (1.16), the weight g dominates in the boundary integral and this dissappears.

Thus, the weighted energy E satisfies

$$\frac{dE}{dt} \leqq \frac{\alpha^2 E}{4},$$

and by integration

$$E(t) \leqq E(0).\exp\left(\frac{\alpha^2 t}{4}\right).$$

Since $E(0) = 0$, it follows that $E(t)\equiv 0$, for every $t>0$, and then $u \equiv 0$.

The method just described is due to Galdi & Rionero [41].

It would appear that the idea of the weighted energy method first appeared in calculations involving L^2 estimates which were employed to establish uniqueness results for elliptic and hyperbolic equations and unique continuation for elliptic equations, see Aronszajn [2], Cordes [21,22], Friedrichs [36], Pederson [148] and Protter [151-153]. In fact, Protter [151] p.110 attributes the basic idea to K.O.Friedrichs, as does Morawetz [125] who refers to "Friedrichs' a,b,c-method". Indeed, Friedrichs [36] employs a weighted energy argument, although his paper is also valuable at this point as it contains several references to early work with energy methods. However, Protter took these ideas much further and in a series of papers published during the years 1960-1963 he laid down a concise and elegant method of approach to tackling problems of uniqueness and unique continuation for several classes of partial differential equations and differential inequalities, including equations of the elliptic, parabolic and hyperbolic type, Protter [153] - [156]; see also Lees & Protter [107], Lees [106], Cohen & Lees [18]. The type of problem studied is exemplified by the following result of Lees & Protter [107], p.373. Let $B(0,R)$ be the ball, centre 0, radius R, in R^n, and let $S(0,R)$ be its surface. For any $\tau(>0)$ define $D = B(0,R)\times[-\tau,0]$, $\partial D = S(0,R)\times[-\tau,0]$, $\bar{D} = D\cup\partial D$. Further, define the parabolic operator $L\equiv A-\partial/\partial t$, where $A \equiv a_{ij}\partial^2/\partial x_i\partial x_j$ is an elliptic operator on $\bar{D}$, which for positive constants m, M satisfies the restrictions,

$$m\xi_i\xi_i \leq a_{ij}\xi_i\xi_j \leq M\xi_i\xi_i\ , \quad \text{for all } \xi_i \text{ and for each } x,t\in\bar{D}.$$

Let v satisfy the differential inequality

$$(Lv)^2 \leqq c_1 v^2 + c_2 \frac{\partial v}{\partial x_i}\frac{\partial v}{\partial x_i} ,$$

in D, with c_1, c_2 positive constants. If $v \equiv 0$ on ∂D and

$$\lim_{t \to 0^-} \int_{B(0,R)} \frac{\partial v}{\partial x_i}\frac{\partial v}{\partial x_i}\, dx = 0,$$

then $v \equiv 0$ in $\bar{D}$. In particular, one consequence of this result is a backward uniqueness theorem for the heat equation, one of the problems raised earlier in the introduction (question (1)).

The weighted energy idea was introduced into continuum mechanics by Serrin [169] who extended the backward uniqueness theorem of Lees & Protter [107] to establish a similar backward uniqueness result for the Navier-Stokes equations. This result and related ones of Edmunds [26,27] are discussed more fully in §11.

Protter, together with Murray (see [131,132,134]) in the early 1970's extended the earlier work and established various decay and unique continuation results for second order equations and differential inequalities and more recently in Protter [157], Murray & Protter [135] and Murray [133] a means was devised of extending these arguments to produce similar types of result for ultrahyperbolic equations. Indeed, Protter's influence on the weighted energy method and its uses can hardly be overestimated. It was the papers of the early 1970's together with the logarithmic convexity work of Knops & Payne [91] which stimulated a series of papers in which growth results pertaining to instability for solutions to differential equations were obtained, including the equations of linear elastodynamics and those for a mixture of linear elastic solids, see Hills & Knops [66], Levine [114] and Straughan [174,176]. More general growth and nonexistence results in nonlinear continuum mechanics were obtained via a weighted energy approach by Hills & Knops [67].

Although the work of Protter and his co-workers contains various types of weight function, including those which depend on both space and time, the basic idea in the weighted energy method has clearly been rediscovered independently by Morawetz [125-129], Strauss [178,179] and Galdi & Rionero [41-46,163,164]. Since the work of Morawetz [125-130] is lucidly reviewed in [129] we do not include detailed descriptions of her results which use weighted energy arguments, although the final chapter does describe similar ideas. The papers of Morawetz [127], Strauss [179], Cooper & Strauss [19], Glassey [56], Glassey & Strauss [58,59] and Marshall, Strauss & Wainger [123] obtain entirely different results to those already mentioned, and are notably concerned with equations which occur in relativistic theories. The work of Galdi & Rionero, on the other hand, concentrates on the Navier-Stokes equations, and similar equations of fluid mechanics, but in a very novel way in that their aim is to employ a weight function primarily so as to allow the solution to have strong spatial growth at infinity in unbounded domain problems.

One of the objectives of this article is to bring together the various results which have been obtained by different writers, all of which possess the unifying feature of weighted integral estimates, although the weight functions are of necessity very different, and to indicate the various types of problems to which weighted energy methods have been applied.

The remainder of this work describes several approaches for proving nonexistence of solutions to nonlinear differential equations and also systematically studies the weighted energy method. Roughly, the article is in two parts; §§2-8 are concerned with nonexistence whereas §§9-20 deal with weighted energy. Inevitably, however, there is overlap in various places since, for example, one of the approaches to nonexistence is via a weighted energy technique.

We shall employ standard notation from partial differential equations and continuum mechanics. In particular, we shall frequently use Cartesian indicial notation, in conjunction with the (Einstein) summation convention; eg. if $\underset{\sim}{u}$ is an n-vector defined on some domain in R^n,

$$u_j u_{i,j} = u_1 \frac{\partial u_i}{\partial x_1} + u_2 \frac{\partial u_i}{\partial x_2} + \ldots + u_n \frac{\partial u_i}{\partial x_n}, \quad i=1,\ldots,n.$$

Also, we shall use Ω to denote a domain in R^n (usually in applications n=3). If Ω is bounded its boundary will be denoted by Γ whereas if Ω is a region exterior to a compact set in R^n its boundary will be denoted by Σ. By $|u|$ we shall mean the absolute value of $\underset{\sim}{u}$, i.e.

$$|u| = (u_i u_i)^{\frac{1}{2}}.$$

Other notation, specific to later sections, is introduced as required.

We make frequent reference to a weighted arithmetic-geometric mean inequality, referred to as inequality (AG), which is included here for convenience.

For $\underset{\sim}{a},\underset{\sim}{b}$ n-vectors, n=1,2,..., with $n<\infty$, and for any $\alpha(>0)$,

$$a_i b_i \leq \frac{1}{2\alpha}|a|^2 + \frac{\alpha}{2}|b|^2. \qquad \text{(AG)}$$

The emphasis in each § is on explaining the methods employed and the relevance of the results and to this end we consider only classical solutions to the appropriate differential equations. By this we shall mean a function which satisfies the differential equation(s) identically together with any associated boundary and initial conditions and which possesses enough differentiability for each of the terms appearing in the differential equations to be themselves continuous functions. In most of the problems discussed here this requirement may be relaxed; however, inclusion of these details would probably obscure the simplicity of the methods.

2 Concavity and nonexistence

We shall use as a model the problem

$$\frac{\partial^2 u}{\partial t^2} = \Delta u + f(u), \quad \text{in } \Omega\times[0,T),$$
$$u = 0, \quad \text{on } \Gamma\times[0,T), \tag{2.1}$$
$$u(x,0) = u_o(x),\ u_t(x,0) = v_o(x),$$

where Ω is a bounded domain in R^n, Δ is the Laplacian operator and $f(u)$ is a nonlinear function with $f(0)=0$ and an associated potential functional $G(u)$, i.e.

$$G(u) = \int_\Omega\int_0^u f(s)dsdx, \tag{2.2}$$

so that

$$\frac{dG}{dt}(u) = \int_\Omega u_t f(u)dx. \tag{2.3}$$

Let $||.||$ and $<\ ,\ >$ be the norm and inner product on $L^2(\Omega)$, i.e.

$$<u,v> = \int_\Omega uvdx, \quad ||u||^2 = \int_\Omega u^2dx.$$

The concavity technique was introduced by Levine [109] and has been exploited in various areas, see [65,77,85-89,96,109-112, 116,117,120,139,146,147,171,173], some of these references being to nonlinear elasticity and other nonlinear models in continuum mechanics.

Basically, the idea is to produce a function F defined on the set of solutions u and to show $F^{-\alpha}$ is a concave function of t for some $\alpha>0$. To this end note that

$$(F^{-\alpha})'' = -\alpha F^{-\alpha-2}[FF'' - (\alpha+1)(F')^2],$$

so we have to establish that

$$FF'' - (\alpha+1)(F')^2 \geqq 0. \qquad (2.4)$$

For the problem (2.1), define (see [110])

$$F(t) = ||u||^2 + \beta(t+t_o)^2, \qquad (2.5)$$

for β, t_o constants to be selected, $\beta, t_o \geqq 0$.

By multiplying (2.1) by u_t and integrating, using also the boundary conditions, we obtain the energy equation

$$E(t) \equiv \tfrac{1}{2}\,||u_t||^2 + \tfrac{1}{2}\,||\text{grad}u||^2 - G(u) = E(0). \qquad (2.6)$$

Furthermore,

$$F'(t) = 2\langle u,u_t\rangle + 2\beta(t+t_o),$$

$$F''(t) = 2\,||u_t||^2 + 2\langle u,u_{tt}\rangle + 2\beta,$$

$$= 4(1+\alpha)\,||u_t||^2 - 2(1+2\alpha)\,||u_t||^2 + 2\langle u,u_{tt}\rangle + 2\beta.$$

Next, substitute for the second term in $||u_t||^2$ from (2.6) and substitute for u_{tt} from (2.1) to find

$$F'' = 4(1+\alpha)\,||u_t||^2 + 2(1+2\alpha)\,||\text{grad}u||^2 - 4(1+2\alpha)G(u)$$

$$-4(1+2\alpha)E(0) + 2\langle u,\Delta u\rangle + 2\langle u,f(u)\rangle + 2\beta. \qquad (2.7)$$

But $\langle u,\Delta u\rangle = -\|\text{grad}u\|^2$, and so when we form the expression in (2.4) we have

$$FF''-(1+\alpha)(F')^2 = -4(1+\alpha)[\langle u,u_t\rangle+\beta(t+t_o)]^2$$
$$+[\|u\|^2+\beta(t+t_o)^2]\{4(1+\alpha)\|u_t\|^2+4(1+\alpha)\beta$$
$$-2(1+2\alpha)\beta-4(1+2\alpha)E(0)+4\alpha\|\text{grad}u\|^2$$
$$+2\langle u,f\rangle - 4(1+2\alpha)G(u)\}$$

$$= 4(1+\alpha)S^2 - 2(1+2\alpha)[\beta+2E(0)]F + 4\alpha\|\text{grad}u\|^2 F$$
$$+ 2F[\langle u,f\rangle - 2(1+2\alpha)G(u)], \tag{2.8}$$

where $S^2 = -[\langle u,u_t\rangle+\beta(t+t_o)]^2+[\|u\|^2+\beta(t+t_o)^2][\|u_t\|^2+\beta] \geq 0$ by the Cauchy-Schwarz inequality.

The first and third terms in (2.8) are non-negative and so to satisfy (2.4) we have to ensure the second and fourth terms are likewise non-negative. Thus, the class of nonlinearities is restricted by the condition

$$\langle u,f\rangle - 2(1+2\alpha)G(u) \geq 0. \tag{2.9}$$

This is a fairly wide class, however, and contains all powers of u greater than 1, eg. if $f(u) = u^r$, $r = 2,3,\ldots$, then

$$\langle u,f\rangle = \int_\Omega u^{r+1}dx, \qquad G(u) = \frac{1}{(r+1)}\int_\Omega u^{r+1}dx,$$

and so $\alpha = (r-1)/4 > 0$. Other types of nonlinearity are given in [110].

<u>Theorem 2.1</u> Suppose f and G satisfy (2.9). Then a global solution to (2.1) does not exist in the cases:

a. $E(0) < 0$;   b. $E(0) = 0$ and $\langle u_o, v_o \rangle > 0$.

Proof. For case (a) choose $\beta = -2E(0) > 0$ and then from (2.8) $FF'' - (1+\alpha)(F')^2 \geqq 0$. Therefore, $(F^{-\alpha})'' \leqq 0$. Integrate this twice to see that

$$(F^{-\alpha}(t))' \leqq -\alpha F^{-\alpha-1}(0)F'(0),$$

$$F^{-\alpha}(t) \leqq F^{-\alpha}(0) - \alpha F^{-\alpha-1}(0)F'(0)t,$$

from which it then follows that

$$F^{\alpha}(t) \geqq \frac{F^{\alpha}(0)}{1-\alpha t(F'(0)/F(0))}. \qquad (2.10)$$

Now $F'(0) = 2\langle u_o, v_o \rangle + 2\beta t_o = 2\langle u_o, v_o \rangle - 4E(0)t_o$. Pick t_o so that $F'(0) > 0$ then the right hand side of (2.10) blows up at $t = F(0)/\alpha F'(0)$. Therefore, the function $F(t)$ must blow up at some time less than or equal to this value or must simply cease to exist for some other reason.

For case (b) choose $\beta = 0$. Then $F'(0) = 2\langle u_o, v_o \rangle > 0$ by hypothesis. Similar arguments establish the theorem.

The proof of the theorem has a nice geometrical interpretation as shown.

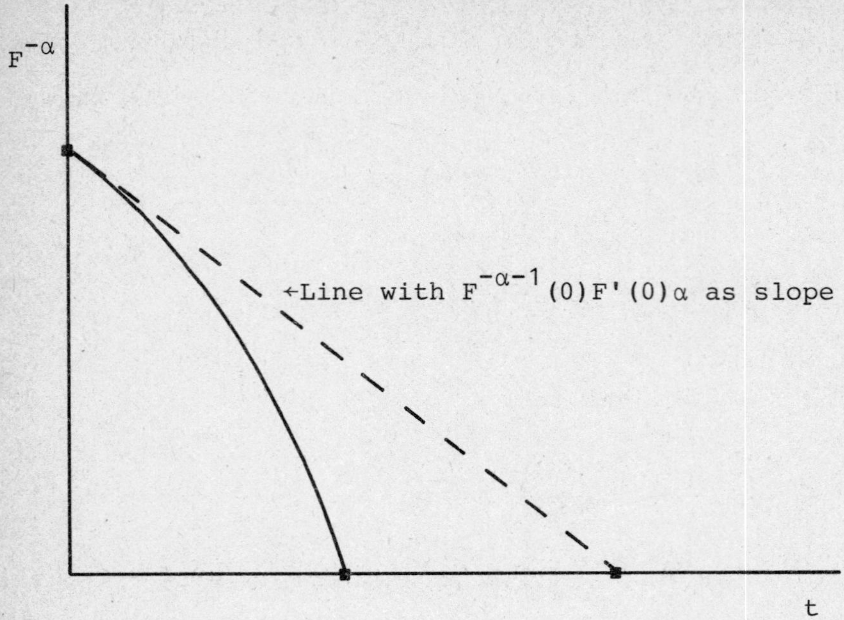

Figure 2.1

Some nonexistence results may be obtained when $E(0)>0$ but the arguments are a little more detailed, see [89,111,112,171].

3 Nonexistence for solutions to the linear wave equation with a nonlinear radiation boundary condition

The same ideas as those of §2 work if instead of (2.1) we consider

$$\frac{\partial^2 u}{\partial t^2} = \Delta u \quad \text{in } \Omega\times(0,T),$$
$$\frac{\partial u}{\partial n} = f(u) \quad \text{on } \Gamma\times[0,T), \tag{3.1}$$
$$u(x,0) = u_o(x), \quad u_t(x,0) = v_o(x),$$

where $\partial u/\partial n$ is the outward normal derivative and f is some non-linear function. Problems of this (and more general) type were first considered by Levine & Payne [116,117].

By multiplying the differential equation by u_t and integrating over Ω we find

$$\langle u_t, u_{tt}\rangle - \langle u_t, \Delta u\rangle = 0.$$

But

$$-\langle u_t, \Delta u\rangle = \tfrac{1}{2}\frac{d}{dt}\,||\text{grad}u||^2 - \int_\Gamma u_t \frac{\partial u}{\partial n} dS$$
$$= \tfrac{1}{2}\frac{d}{dt}\,||\text{grad}u||^2 - \int_\Gamma u_t f(u)\, dS.$$

Hence, if g(u) is now a potential functional for f in the boundary sense, i.e. $\int_\Gamma g(u)ds = \int_\Gamma\int_0^u f(a)da dS$, the energy equation for (3.1) is

$$E(t) \equiv \tfrac{1}{2}\,||u_t||^2 + \tfrac{1}{2}\,||\text{grad}u||^2 - \int_\Gamma g(u)dS = E(0). \tag{3.2}$$

To obtain a nonexistence result F(t) is again defined as

$$F(t) = \|u\|^2 + \beta(t+t_o)^2. \tag{3.3}$$

Then analogous to (2.7) one finds

$$F'' = 4(1+\alpha)\|u_t\|^2 + 2(1+2\alpha)\|\text{grad}u\|^2 - 4(1+2\alpha)\int_\Gamma g(u)dS$$

$$- 4(1+2\alpha)E(0) + 2\langle u,\Delta u\rangle + 2\beta. \tag{3.4}$$

Now, however,

$$\langle u,\Delta u\rangle = -\|\text{grad}u\|^2 + \int_\Gamma u\frac{\partial u}{\partial n}dS$$

$$= -\|\text{grad}u\|^2 + \int_\Gamma uf(u)dS. \tag{3.5}$$

Hence, combining (3.4) and (3.5) we find instead of (2.8)

$$FF''-(1+\alpha)(F')^2 = 4(1+\alpha)S^2 - 2(1+2\alpha)[\beta+2E(0)]F$$

$$+ 4\alpha\|\text{grad}u\|^2 F + 2F\{\int_\Gamma uf(u)dS - 2(1+2\alpha)\int_\Gamma g(u)dS\}, \tag{3.6}$$

and so if we replace (2.9) by the condition

$$\int_\Gamma [uf(u) - 2(1+2\alpha)g(u)]dS \geqq 0, \tag{3.7}$$

an exactly similar theorem to theorem 2.1 holds.

4 The convergent integral method

This method is the one which appears to have been most widely used and essentially consists of assuming a solution to some problem exists for all values of a certain parameter and then obtaining a contradiction by showing that the parameter is bounded above by an integral which converges, in a similar fashion to (1.2) in the simple example in the introduction.

As far as I am aware the method was first used by Wittich [192] to study the nonlinear elliptic equation $\Delta u = e^u$. Further extensions to elliptic equations and related results, topics we do not cover, are given in [62,82-84,121,148,166,184]; applications of the method to other areas are contained in [3,31,37-39, 54,55,94,96,105,113,118,119,139].

Let us reconsider problem (2.1). We follow the method in the form given by Glassey [54] and Levine [113]. Let ϕ be the first eigenfunction of the membrane problem for Ω, i.e. ϕ satisfies

$$\Delta\phi + \lambda_1\phi = 0 \quad \text{in } \Omega, \qquad (4.1)$$
$$\phi = 0 \quad \text{on } \Gamma.$$

By the Courant nodal line theorem we know that ϕ is one-signed and so we may select $\phi>0$ in Ω.

Define now $F(t)$ by

$$F(t) = \int_\Omega \phi(x)u(x,t)dx, \qquad (4.2)$$

where u is a solution to (2.1) and where we assume the nonlinear function f in (2.1) is convex in u, although we no longer require $f(0) = 0$. Let us also suppose ϕ is normalized, i.e. $\int_\Omega \phi dx = 1$. Moreover, let H be an indefinite integral of f, i.e. $H'(p) = f(p)$, p real, and suppose that there exists $s>0$ such that

$$2H(p) - \lambda_1 p^2 \quad \text{is non-decreasing on } [s,\infty). \tag{4.3}$$

[To fix ideas consider $f(p) = e^p$, then $H(p) = e^p$ and we can obviously find an s to satisfy (4.3).]

Theorem 4.1 If f is convex, (4.3) holds and

a. $F(0) \geqq s$; b. $F'(0) > 0$;

c. $$\int_{F(0)}^{\infty} \frac{dF}{\sqrt{[(F'(0))^2+\lambda_1 F^2(0)-2H(F(0))-\lambda_1 F^2+2H(F)]}} < \infty;$$

then the solution to (2.1) does not exist for all time.

Proof. From (4.2),

$$\begin{aligned} F''(t) &= \int_\Omega \phi u_{tt} dx \\ &= \int_\Omega \phi \Delta u dx + \int_\Omega \phi f(u) dx \\ &= \int_\Omega u \Delta\phi dx + \int_\Omega \phi f(u) dx, \end{aligned}$$

where (2.1) has been used to substitute for u_{tt} and we have then integrated by parts the Δu term. Since ϕ satisfies (4.1), we may write

$$\begin{aligned} F'' &= -\lambda_1 \int_\Omega \phi u dx + \int_\Omega \phi f(u) dx, \\ &= -\lambda_1 F + \int_\Omega \phi f(u) dx. \end{aligned} \tag{4.4}$$

The nonlinear term is handled with the aid of Jensen's inequality which for convex f may be written

$$f\left(\int_\Omega \phi u dx\right) \leqq \int_\Omega \phi f(u) dx \Big/ \int_\Omega \phi dx.$$

Using this in (4.4), we find (because ϕ is normalized)

$$F'' \geqq -\lambda_1 F + f(F). \tag{4.5}$$

Now, $F'(0)>0$ and so either $F'>0$ for all t or $F'(\eta)=0$ for some $\eta < \infty$. Suppose the second case. Then, for $0 \leqq t < \eta$, multiply (4.5) by F' and integrate to obtain

$$\tfrac{1}{2}[F'(t)]^2 \geqq \tfrac{1}{2}[F'(0)]^2 - \tfrac{1}{2}\lambda_1 F^2(t) + H(F(t))$$
$$+\tfrac{1}{2}\lambda_1 F^2(0) - H(F(0)). \tag{4.6}$$

F is increasing on $[0,\eta)$ and so

$$-\tfrac{1}{2}\lambda_1 F^2(t)+H[F(t)]+\tfrac{1}{2}\lambda_1 F^2(0)-H[F(0)] \geqq 0,$$

since $F(0) \geqq s$; therefore, for $0 \leqq t < \eta$,

$$F'(t) \geqq F'(0) > 0.$$

By continuity, therefore, $F'(\eta)$ cannot be zero; contradiction. It follows that $F'>0$ for every t for which the solution exists and so (4.6) holds for the same range of t values. Since both sides of (4.6) are now non-negative we may take square roots, separate variables and integrate to see that

$$t \leqq \int_{F(0)}^{F(t)} \frac{dF}{\surd\{[F'(0)]^2+\lambda_1 F^2(0)-2H[F(0)]-\lambda_1 F^2+2H(F)\}}$$

$$\leqq \int_{F(0)}^{\infty} \frac{dF}{\surd\{[F'(0)]^2+\lambda_1 F^2(0)-2H[F(0)]-\lambda_1 F^2+2H(F)\}} \tag{4.7}$$

$< \infty$, by hypothesis.

Therefore, the solution cannot exist for all time. In fact, an upper bound for the interval of existence is the number given by the integral in (4.7).

The concavity method and the one just described do not necessarily handle the same type of problems. For example, the initial data may often belong to a larger class with the concavity method and the concavity arguments are applicable to a wider range of differential equations. However, the convergent integral method does have some advantages, namely:

a. No energy equation is used and so there is no need to restrict $E(0)$.

b. The types of nonlinearities include functions such as e^u, $u\exp(u^2)$ which are not so convenient for the concavity method.

c. The nonlinearity may depend only on u_t instead of u, see [54,113].

It is a good point to raise a question regarding the dependence of the escape time for the solution to (2.1) upon the geometry. That is, let us fix $F(0)$ and $F'(0)$ as defined in this § and fix also the volume, V, of Ω. Given that there is nonexistence, which shape of Ω, for fixed V, is the most and least stable; in other words, if there is blow-up, which $F(t)$ blows up fastest and which is slowest. I am aware of no results of this type whatsoever for (2.1). By a naive argument based on the above results we may conjecture that the sphere will be the least stable; for, $\lambda_1 \geq \lambda_{1\text{sphere}}$, for fixed V (see Payne [144]) and so the integral in (4.7) is smallest for Ω an n-sphere.

For the equivalent parabolic problem for (2.1), Bandle [4] has established that the n-sphere is the least stable.

5 The convergent integral method of Levine and Payne

So far we have concentrated on techniques which require the linear spatial operator, in our case Δ, to be a negative operator, i.e. $\langle u,\Delta u\rangle = -\|\text{grad}u\|^2 \leqq 0$. A method introduced by Levine & Payne [118,119] allows, among other things, the operator to be sign indefinite.

In accordance with the approach of the earlier §§ we shall illustrate the technique by means of a simple example. An abstract equation which covers many situations in partial differential equations is treated specifically in [118] while nonlinear elastodynamics is covered in [119]. Let us consider now,

$$
\begin{aligned}
&\frac{\partial^2 u}{\partial t^2} = \frac{\partial^2 u}{\partial x \partial y} + f(u), \qquad && \text{in } \Omega\times[0,T),\\
&u = 0, && \text{on } \Gamma, \qquad\qquad (5.1)\\
&u(x,0) = u_o(x), \quad u_t(x,0) = v_o(x),
\end{aligned}
$$

where now Ω is a bounded domain in R^2. We first observe that the techniques of §§2 and 4 break down for (5.1) because

$$-\langle u,u_{xy}\rangle = \langle u_x,u_y\rangle,$$

which may be either positive or negative. Moreover, to apply the ideas of chapter 4 we would need a positive eigenfunction for the problem $\phi_{xy}+\lambda\phi=0$ and the Volterra equation $\phi=-\lambda\int\int\phi$ has only the zero solution.

Instead, we shall review the presentation of [118], and so introduce

$$J(t) = \|u(t)\|^2 . \tag{5.2}$$

Then, as usual,

$$J'' = 2\langle u,u_{tt}\rangle + 2\|u_t\|^2$$
$$= 2\|u_t\|^2 - 2\langle u_x,u_y\rangle + 2\langle u,f\rangle.$$

It is necessary to remove the $\langle u_x,u_y\rangle$ term, and so we derive the energy equation for (5.1),

$$E(t) \equiv \tfrac{1}{2}\|u_t\|^2 + \tfrac{1}{2}\langle u_x,u_y\rangle - G(u) = E(0), \tag{5.3}$$

where G(u) is the potential introduced in §2.

We substitute for $\langle u_x,u_y\rangle$ from (5.3) to find

$$J'' = 4\|u_t\|^2 - 4G(u) - 4E(0) + 2\langle u,f\rangle.$$

Now, multiply by J and use the fact that

$$4\|u\|^2\|u_t\|^2 \geqq 4\langle u,u_t\rangle^2 = (J')^2 \quad \text{(Cauchy-Schwarz)}$$

to see that

$$JJ'' \geqq (J')^2 - 4E(0)J + 2[\langle u,f\rangle - 2G(u)]J.$$

At this stage it is necessary to say something about the nonlinearity f. Suppose

$$\langle u,f\rangle - 2G(u) \geqq J^2H'(J), \tag{5.4}$$

for a function H to be described more fully later. Then, introduce this in the expression for JJ" and divide by J^2 and we obtain

$$\frac{d}{dt}\left(\frac{J'}{J}\right) \geq -4\frac{E(0)}{J} + 2JH'(J) \ . \tag{5.5}$$

As in §4, if $J'(0)>0$ either $J'>0$ on $[0,\eta)$ with $J'(\eta)=0$ or $J'>0$ for all t. Suppose $J'(\eta)=0$. Then, on $[0,\eta)$ we multiply (5.5) by J'/J and integrate to find

$$\frac{(J')^2}{J} \geq \left[\left(\frac{J'(0)}{J(0)}\right)^2 - 8\frac{E(0)}{J(0)}\right]J + 8E(0) + 4J[H(J)-H(J(0))]. \tag{5.6}$$

Define the right hand side of (5.6) by $R(J)$.

To proceed we impose the conditions

$$\begin{array}{ll} \text{(i)} & J(0) > 0 \quad \text{(already implicitly assumed)}, \\ \text{(ii)} & J'(0) > 0, \\ \text{(iii)} & JH(J) \text{ is convex in } J, \\ \text{(iv)} & \left(\frac{J'(0)}{J(0)}\right)^2 - \frac{8E(0)}{J(0)} + 4J(0)H'(J(0)) \geq 0. \end{array} \tag{5.7}$$

Now,

$$\frac{dR}{dJ} = \{[J'(0)/J(0)]^2 - 8E(0)/J(0) + 4J(0)H'(J(0))\}$$

$$+4[H(J)-H(J(0))] + 4[JH'(J)-J(0)H'(J(0))].$$

The first term is non-negative by (5.7)(iv). Moreover, $M=JH$ is convex and so $d^2M/dJ^2 \geq 0$ and, therefore,

$$\frac{dM}{dJ}(J(t)) \geq \frac{dM}{dJ}(J(0)),$$

since J is increasing on $[0,\eta)$. $dM/dJ = H+JH'$ and so $dR/dJ \geq 0$. Hence $R(J(t)) \geq R(J(0))$ on $[0,\eta)$. But $R(J(0)) = [J'(0)]^2/J(0) > 0$ and so by continuity $R(J(\eta)) \neq 0$ which implies from (5.6) that $J'(\eta) \neq 0$, contradiction. Therefore, $J'>0$ for each t and likewise (5.6) holds for all t. Since also $R(J)>0$ we may rearrange (5.6), take square roots and separate variables to see that

$$t \leq \int_{J(0)}^{\infty} \frac{dJ}{\sqrt{P(J)}} \tag{5.8}$$

where

$$P(J) = \left[\left(\frac{J'(0)}{J(0)}\right)^2 - 8\frac{E(0)}{J(0)}\right]J^2 + 8E(0)J + 4J^2[H(J) - H(J(0))].$$

If the right hand side of (5.8), which we denote by T#, is finite then the solution cannot exist for all time.

Hence, provided (5.4), (5.7) and (5.8) are met, a solution to (5.1) cannot exist beyond T#.

Remarks.

1. No explicit condition on the sign of E(0) is made, although the size of E(0) is restricted by (5.7)(iv).
2. Instability and nonexistence results for Cauchy elasticity, along similar lines to those of this §, are given in [94,96,174].

6 Various problems in which nonexistence occurs and solution structure

In §§2-5 we have seen several methods for establishing nonexistence and although they are very useful there are several drawbacks. Among these we mention there are some problems in applied mathematics which require very special techniques and are not amenable by the general methods outlined earlier. Another point is that no indication is given as to whether "blow-up" actually occurs, nor is a lower bound obtained for the time of existence. Finally, owing to the very general nature of the concavity and convergent integral approaches which are applicable to a very wide class of equations, no information is obtained regarding the solution structure, by which we mean that if blow-up occurs how is the shape of the solution behaving near the break up time. We turn in this § to address questions of the type just outlined. But, as work in this direction is of recent origin we are only able to pick a selection of the many new avenues being explored. Nevertheless, it is hoped the reader will see the diverse nature of questions that can be raised in connection with nonexistence.

We begin with a study of a nonlinear Schrödinger equation analysed by Glassey [55].

The initial-value problem is

$$iu_t = a_{ij}u_{,ij} + f(|u|^2)u, \quad x \in R^n,\ t > 0, \tag{6.1}$$

$$u(x,0) = \phi(x).$$

Here $u=u_1(x)+iu_2(x)$, f is a nonlinear term to be described below and a_{ij} is a symmetric n×n matrix with constant entries (Glassey [55] studies the case $a_{ij}=\delta_{ij}$; however, the generalization is convenient as it allows us to study a problem of Jeffrey[70]). Throughout, we shall suppose that u and ϕ decay sufficiently rapidly as $|x|\to\infty$ in order that the integrals introduced are meaningful, and let $\|.\|$ denote the norm on $L^2(R^n)$.

The instability and nonexistence results rely on the following identities.

<u>Lemma 6.1</u> (Glassey [55]).

$$\|u\|^2 = \|\phi\|^2, \tag{6.2}$$

$$\frac{d}{dt}\int |x|^2|u|^2dx = -4\int x_r a_{rs}(u_1u_{2,s}-u_2u_{1,s})dx, \tag{6.3}$$

$$\int a_{rs}(u_{1,r}u_{1,s}+u_{2,r}u_{2,s})dx-\int g(|u|^2)dx = E_o, \tag{6.4}$$

$$\frac{d}{dt}\left(\mathrm{Im}\int x_m\bar{u}u_{,m}dx\right) = -2\int a_{ij}u_{,i}\bar{u}_{,j}dx + n\int[f(|u|^2)|u|^2-g(|u|^2)]dx, \tag{6.5}$$

where E_o is the left hand side of (6.4) with u replaced by ϕ, g is the potential for f, i.e. $g(s) = \int_0^s f(a)da$, $\bar{u}=u_1-iu_2$, and all integrals are over R^n.

Proof. The establishment of (6.2) and (6.3) proceeds by multiplying $(6.1)_1$ by $2\bar{u}$ and then taking the imaginary part of the resulting expression, to obtain

$$\frac{\partial}{\partial t}|u|^2 = 2[a_{rs}(u_1u_{2,s}-u_2u_{1,s})]_{,r}. \tag{6.6}$$

For u_α and $u_{\alpha,s}$ which decay sufficiently rapidly as $|x|\to\infty$, (6.6) leads immediately to (6.2).

Expression (6.3) follows after multiplying (6.6) by $|x|^2$, integrating over R^n and integrating by parts the right hand side.

To derive (6.4), multiply $(6.1)_1$ by $2\bar{u}_t$. Then, integrate the real part of the corresponding identity over R^n and after an integration by parts on the a_{rs} term we find

$$0 = -\frac{d}{dt}\int a_{rs}(u_{1,s}u_{1,r}+u_{2,s}u_{2,r})dx+\int f(|u|^2)\frac{\partial}{\partial t}|u|^2dx. \tag{6.7}$$

Since g is the potential for f the last expression is just

$$\frac{d}{dt}\int g(|u|^2)dx,$$

and (6.4) follows by integration of (6.7).

It remains to prove (6.5) and for this we multiply $(6.1)_1$ by $2x_m\bar{u}_{,m}$, i.e.

$$2ix_m\bar{u}_{,m}u_t = 2x_m\bar{u}_{,m}(a_{rs}u_{,s})_{,r} + 2x_m fu\bar{u}_{,m}. \tag{6.8}$$

Observe now that

$$\begin{aligned}\mathrm{Re}\int 2ix_m\bar{u}_{,m}u_t dx &= \mathrm{Re}\int ix_m[(\bar{u}_{,m}u)_{,t}-(u\bar{u}_{,t})_{,m}]dx \\ &= \frac{d}{dt}\left(\mathrm{Re}\int ix_m u\bar{u}_{,m}dx\right)+n\mathrm{Re}\int i\bar{u}_{,t}udx \\ = \frac{d}{dt}\left(\mathrm{Re}\int ix_m\bar{u}_{,m}udx\right) &+ n\int[a_{rs}u_{,r}\bar{u}_{,s}-f(|u|^2)|u|^2]dx,\end{aligned}$$

where in the last term $(6.1)_1$ has been reused. Finally, the first term on the right is rearranged and we see that

$$\begin{aligned}\mathrm{Re}\left(\int 2ix_m\bar{u}_{,m}u_t dx\right) &= \frac{d}{dt}\left(\mathrm{Im}\int x_m\bar{u}u_{,m}dx\right) \\ &+ n\int[a_{rs}u_{,r}\bar{u}_{,s}-f(|u|^2)|u|^2]dx.\end{aligned} \tag{6.9}$$

By writing u, $\bar{u}$ in terms of u_1, u_2 and integrating by parts it follows easily that

$$\mathrm{Re}\int 2x_m\bar{u}_{,m}(a_{rs}u_{,s})_{,r}dx=(n-2)\int a_{rs}[u_{1,s}u_{1,r}+u_{2,s}u_{2,r}]dx. \tag{6.10}$$

Also, similar calculations establish that

$$\mathrm{Re}\int 2x_m f\bar{u}_{,m}u dx = -n\int g(|u|^2)dx. \tag{6.11}$$

We next integrate the real part of (6.8) over R^n and (6.5) follows on substituting (6.9)-(6.11) in the resulting equation.

We are now able to describe the main nonexistence result of Glassey [55]. Suppose $a_{ij}=\delta_{ij}$ and f and g are such that

$$sf(s) - g(s) \geqq \alpha_n g(s), \tag{6.12}$$

for all $s \geqq 0$ and a constant $\alpha_n > 2/n$.

Theorem 6.1 (Glassey [55]). Let u be a classical solution to (6.1) and suppose

$$E_o \leqq 0, \tag{6.13}$$

$$\mathrm{Im}\int x_m\bar{\phi}\phi_{,m}dx > 0. \tag{6.14}$$

Then u cannot exist for all time. In particular, if there are constants $k,\sigma > 0$ such that

$$g(s) \leqq ks^{1+\sigma}, \quad \text{for all } s > 0, \tag{6.15}$$

then $\sup_{R^n}|u|$ is bounded below by a function of t which blows up in finite time.

Proof. Assuming existence for t in a neighbourhood of 0 define y(t) by

$$y(t) = \mathrm{Im}\int x_m \bar{u} u_{,m} dx. \tag{6.16}$$

From (6.5)

$$y'(t) = -2\,\|\nabla u\|^2 + n\int [|u|^2 f(|u|^2) - g(|u|^2)]dx$$

$$\geq -2\,\|\nabla u\|^2 + n\alpha_n \int g(|u|^2)dx, \tag{6.17}$$

where (6.12) has been employed. Substitute now for the g term from (6.4) to arrive at

$$y' \geq (n\alpha_n - 2)\,\|\nabla u\|^2 - n\alpha_n E_o$$

$$\geq (n\alpha_n - 2)\,\|\nabla u\|^2, \tag{6.18}$$

since $E_o \leq 0$. Observe that $n\alpha_n - 2 > 0$.

If we write $a_{ij} = \delta_{ij}$ in (6.3) we find

$$\frac{d}{dt}\int |x|^2 |u|^2 dx = -4y(t). \tag{6.19}$$

By (6.18) y is increasing and since from (6.14) $y(0)>0$, $y(t)>0$, hence from (6.19),

$$\int |x|^2 |u|^2 dx \leq \int |x|^2 |\phi|^2 dx \overset{\text{def}}{=} B_o < \infty. \tag{6.20}$$

Employing the Cauchy-Schwarz inequality,

$$y^2(t) \leq \int |x|^2 |u|^2 dx.\, \|\nabla u\|^2 \leq B_o\,\|\nabla u\|^2. \tag{6.21}$$

This is now combined with (6.18) to yield

$$y' \geq \frac{(n\alpha_n - 2)}{B_o} y^2, \tag{6.22}$$

which may be integrated to obtain

$$t \leqq B_o/(n\alpha_n-2)y(0) = T_u. \tag{6.23}$$

A contradiction is obtained for t greater than the value T_u and so we conclude a classical solution cannot exist for all time.

For values of t for which the solution exists, (6.22) when integrated and combined with (6.21) gives the following estimate for the growth rate of $||\nabla u||$,

$$||\nabla u|| \geqq B_o^{\frac{1}{2}}y(0)/[B_o-(n\alpha_n-2)ty(0)]. \tag{6.24}$$

It remains to establish a lower bound for $\sup|u|$. Since $E_o \leqq 0$ from (6.4) we have

$$||\nabla u||^2 \leqq \int g(|u|^2)dx \leqq k\int |u|^{2+2\sigma}dx$$

$$\leqq k[\sup_{R^n} |u|]^{2\sigma} ||\phi||^2 ,$$

where (6.2) has been used. From (6.24), the above estimate shows $\sup|u|$ is bounded below by a function which blows up in a finite time.

It is also worth pointing out that if we retain the E_o term in (6.18) then the upper bound T_u may be replaced by

$$T_u = \frac{B_o^{\frac{1}{2}}}{[n\alpha_n(n\alpha_n-2)|E_o|]^{\frac{1}{2}}}.\tan^{-1}\left[\left(\frac{1-2/n\alpha_n}{|E_o|}\right)^{\frac{1}{2}}\frac{y(0)}{B(0)}\right]. \tag{6.25}$$

We return to these estimates later.

Next, we consider a special case of (6.1) not covered in [55]. This was motivated by work of Jeffrey [70] who gave evidence of instability in R^2 for the equation

$$iu_t = u_{xx} - u_{yy} + |u|^2u, \tag{6.26}$$

with periodic initial conditions. We consider, therefore, (6.1) in R^2 with a general symmetric a_{rs} with constant coefficients and with $f(s) = s$ so that $g(s) = s^2/2$. Again, define $y(t)$ as in (6.16). Equation (6.5) leads in this case to

$$y' = -2\int a_{rs}u_{,r}\bar{u}_{,s}dx + \int |u|^4 dx,$$

$$= -2E_o ,$$

after use of (6.4). Hence,

$$y(t) = -2E_o t + y(0). \tag{6.27}$$

If now $E_o<0$, and $a_{ij}=\delta_{ij}$, then we may use (6.21) in (6.27) to see that

$$||\nabla u|| \geqq -2E_o t/B_o^{\frac{1}{2}} + y(0)/B_o^{\frac{1}{2}} , \tag{6.28}$$

from which we conclude instability, albeit only linear growth in t.

In the general a_{rs} case we cannot conclude (6.19) from (6.3) and so no longer have (6.21). However, the result

$$y^2 \leqq \int |x|^2|u|^2dx. ||\nabla u||^2 , \tag{6.29}$$

is still true and so when $E_o < 0$, either

(i) $\int |x|^2|u|^2dx \to \infty$ as $t \to \infty$, or

(ii) there exists $A(<\infty)$ such that $\int |x|^2|u|^2dx \leqq A$ for every $t > 0$.

In the second case (6.29) together with (6.27) show that $||\nabla u|| \to \infty$ as $t \to \infty$ and so the solution is unstable under either situation.

We now consider further the behaviour of a solution to (6.1). Strictly the material described next should be in §7 on decay results. However, as the decay behaviour has a direct bearing on the solution structure to the nonlinear Schrödinger equation it is preferable to include it in the present position.

Suppose $a_{ij}=\delta_{ij}$, then define B by

$$B(t) = \int |x|^2 |u|^2 dx. \tag{6.30}$$

From (6.4), (6.5) and (6.19) we have

$$\frac{dB}{dt} = -4y , \tag{6.31}$$

$$\frac{dy}{dt} = -2 \|\nabla u\|^2 + n\int [|u|^2 f(|u|^2)-g(|u|^2)]dx, \tag{6.32}$$

$$E_o = \|\nabla u\|^2 - \int g(|u|^2)dx, \tag{6.33}$$

and so

$$B'' = 8 \|\nabla u\|^2 - 4n\int [|u|^2 f(|u|^2)-g(|u|^2)]dx. \tag{6.34}$$

Suppose (6.12) holds for a constant α_n such that $\alpha_n \geqq 2/n$. Then (6.34) together with (6.12) yield

$$B'' \leqq 8 \|\nabla u\|^2 - 4n\alpha_n \int g(|u|^2)dx.$$

By substituting for the g term from (6.33), we see that

$$B'' \leqq -4(n\alpha_n-2) \|\nabla u\|^2 + 4n\alpha_n E_o. \tag{6.35}$$

Let us now require the earlier conditions (6.13), (6.14) to hold in addition; here (6.14) is equivalent to $B'(0)<0$. Since $n\alpha_n \geqq 2$ we find B is a concave function in t and so

$$B(t) \leqq B_o + B'(0)t + 2n\alpha_n E_o t^2, \tag{6.36}$$

where $B_o = B(0)$ (cf. (6.20)). This implies that, provided u exists long enough, $B \to 0$ in a finite time. In particular there are two obvious results;

(i) if $E_o \leqq 0$, $B'(0) < 0$ then

$$B(t) \to 0 \quad \text{as } t \to T^- \leqq T_d = -B_o/B'(0); \tag{6.37}$$

(ii) if $E_o < 0$, $B'(0) = 0$ then

$$B(t) \to 0 \quad \text{as } t \to T^- \leqq T_d^* = (-B_o/2\alpha_n E_o)^{\frac{1}{2}}; \tag{6.38}$$

assuming existence to T^- in both cases.

The conditions for (6.37) are exactly those which hold in Glassey's nonexistence theorem (theorem 6.1; except we can allow also $n\alpha_n=2$) and so it is worth investigating the two possible phenomena $B \to 0$, $\sup|u| \to \infty$, which would appear to be able to occur simultaneously. $B \to 0$ means that

$$\int |x|^2 |u|^2 dx \to 0,$$

and remembering the behaviour of a δ-function and its derivatives

$$\int f(x)\delta^{(n)}(x)dx = (-1)^n f^{(n)}(0),$$

it is appealing to conjecture that in this problem the solution becomes a distribution in finite time, possibly a combination of δ and δ'.

If we look at the corresponding escape times, from (6.23) and (6.37), then

$$T_u = \frac{-4B_o}{[n\alpha_n - 2]B'(0)}, \quad T_d = \frac{-B_o}{B'(0)},$$

Unfortunately there does not appear to be much information to be obtained from this, especially as T_u and T_d are only estimates.

To conclude this topic it is worth writing down the condition $n\alpha_n \geqq 2$ for $f(|u|^2) = |u|^{p-1}$. Then, the condition is $p \geqq 1+4/n$. It would be interesting to study the blow up question in this case by either more sophisticated analysis or numerical means to determine the solution structure, to see if δ-function behaviour does occur.

To study solution structure it would appear necessary to abandon the general methods for establishing nonexistence and concentrate on individual equations. For the type of problem envisaged here there appears to be very little work which has investigated solution shape as time evolves and particularly, in the vicinity of the blow-up time when blow-up does occur. There is, however, one very interesting piece of work devoted to such questions; a thesis of Padalino [139]. She studies numerically the equation $u_t=u_{xx}+u^n$, primarily when $n=2$. The spatial region is $0 < x < \pi$ and the boundary condition $u = 0$ is chosen. Although there are many questions left unanswered her work is an encouraging start in an area in which further study should prove rewarding. The numerical methods adopted by Padalino [139] involve explicit and implicit difference schemes and a Galerkin-Taylor series method. Some of her interesting findings are outlined below.

1. The Galerkin method is thought most accurate and this technique indicated a blow-up profile less peaky than the implicit and explicit methods, the latter indicating the most

peaked solution.

2. For the equation

$$u_t = u_{xx} + u^2,$$

with the initial condition

$$u(x,0) = k\sin 2x,$$

strong evidence was obtained that the nodal point initially at $\pi/2$ moved steadily across to $x = \pi$, even when blow-up occurred. However, the solution to

$$u_t = u_{xx} - u^2,$$

subject to the same initial condition was such that the nodal line wandered from $\pi/2$ to 0. For comparison, it was proved that the same initial condition together with the equation

$$u_t = u_{xx} + u^3,$$

leaves the nodal line at $\pi/2$.

3. In general, the escape time obtained from the convergent integral method is sharper than that found from concavity, although both estimates may be very conservative compared with the escape time found numerically.

It is clear that a detailed study of solution structure and the unpredictable nature of the nonlinear equations found in applied mathematics may require new methods to be developed. Hence, in the remainder of this § we describe some special problems and methods which have recently appeared in the literature.

Sperb [170], by establishing a maximum principle for

$$G(u)u_t + H(u),$$

for G, H suitably chosen deduces nonexistence results for

$$u_t = \Delta u + f(u), \quad \text{in } \Omega\times(0,T),$$

$$u = 0, \quad \text{on } \Gamma\times[0,T), \tag{6.39}$$

$$u = u_o(x), \quad \text{when } t = 0.$$

One such result is now quoted.

<u>Theorem 6.2</u> (Sperb [170]). Let u be a smooth positive solution to (6.39) with f such that

$$f(0)=0,\ f'(0)>\lambda_1,\ f''(s)\geqq 0 \text{ for } s\geqq 0,\quad \int_1^\infty f^{-1}(s)ds < \infty.$$

Suppose also $u_o = \phi$, where ϕ is the first eigenfunction in the membrane problem (4.1). Then u(x,t) must become infinite for some time t_c, with $T_* \leqq t_c \leqq T^*$, where

$$T_*=\int_1^\infty f^{-1}(s)ds, \quad T^*=[f'(0)/(f'(0)-\lambda_1)]\int_1^\infty f^{-1}(s)ds.$$

Obviously, the blow-up is pointwise and so the method is much different from those described here. Also, however, it may be applicable to a different type of nonlinearity. eg. $f(u) = \lambda ue^u$, $\lambda>\lambda_1$. A particularly useful thing is that both upper and lower bounds are obtained for t_c. Other lower bounds for the escape time are derived in Padalino [139] and Payne [146].

Another interesting class of problems are those connected with the so-called phenomenon of quenching which arises in the study of electric current transients in polarized ionic conductors. Kawarda [81] investigated the boundary-initial value problem,

$$u_t = u_{xx} + \frac{1}{(1-u)}, \quad 0 < x < L, \ t > 0,$$

$$u = 0, \quad x = 0, L, \quad t > 0, \qquad (6.40)$$

$$u(x,0) = 0, \quad 0 \leqq x \leqq L.$$

He showed that

(i) if $L > 2\sqrt{2}$, then $u(\frac{1}{2}L,t) \to 1$ in finite time;

(ii) if $u(\frac{1}{2}L,t) \to 1$ in finite time, then $u_t(\frac{1}{2}L,t)$ is unbounded in finite time.

When (i) holds, it is said that u quenches in finite time. The breakdown in this case is due to the derivatives of u becoming unbounded rather than the solution itself and so the situation is somewhat analogous to that for nonlinear hyperbolic conservation laws. Acker & Walter [1] and Levine & Montgomery [115] demonstrated that there exists $L_o \cong 1.5307$ such that if $L > L_o$, u quenches in finite time whereas if $L < L_o$, u tends monotonically to the smaller of the two solutions to the stationary problem

$$f''(x) + \frac{1}{1-f} = 0, \quad 0 < x < L,$$

$$f(0) = f(L) = 0.$$

These results essentially relied on a maximum principle.

An analysis of the equivalent problem for the hyperbolic version of (6.40), namely

$$u_{tt} = u_{xx} + 1/(1-u), \quad 0<x<L, \ t>0,$$

$$u = 0, \quad x = 0, L, \qquad (6.41)$$

$$u(x,0) = u_t(x,0) = 0, \quad 0 \leq x \leq L,$$

was initiated by Chang & Levine [15]. In fact, they studied a wider class of nonlinearities, but that quoted serves for this description. By studying various problems related to (6.41) and eventually reducing everything to a convergent integral argument they showed that if $L > L_1 \cong 1.418$ then u quenches in finite time. On the other hand, they established that if $L < 1.238$ then $u \leqq 0.7732$ for all time, so u does not quench for all L, even in infinite time.

To complete this § we review another nonexistence problem in which the fine detail sought necessitates development of a special analysis. The problem concerns the existence of a classical solution to the nonlinear Cauchy problem

$$\begin{aligned} u_{tt} &= \Delta u + |u|^p, \quad x \in R^n,\ t > 0, \\ u(x,0) &= f(x),\ u_t(x,0) = g(x). \end{aligned} \tag{6.42}$$

The question is to see whether there is a critical value of p, depending on n, $p_o(n)$ say, such that if $p > p_o$ there is global existence of "small" solutions but when $1 < p < p_o$ then there is global nonexistence of most solutions. In fact, John [72] has answered the above question affirmatively when n=3; he finds $p_o(3) = 1+\sqrt{2}$. The only restrictions he imposes are that the data be smooth enough and have compact support, although the case $p=p_o$ is not considered. Kato [80] also investigates similar problems while Glassey [57] continues the study of John. Glassey investigates the nonexistence part of the above question for $n \leqq 3$ and shows that nonexistence does occur when $p=p_o$, n=3 and the data are large enough. Precisely, he demonstrates the validity of the following theorem.

Theorem 6.3 (Glassey [57]). Let $T>0$, $n\leq 3$ and $u\in C^2(R^n\times[0,T))$ be a solution to (6.42). A) Assume that

(i) $f,g \in C_0^\infty(R^n)$, $\mathrm{supp}\{f,g\} \subseteq \{|x|\leq k\}$;

(ii) $\int_{R^n} f(x)dx > 0$, $\int_{R^n} g(x)dx > 0$;

(iii) $1 < p < \frac{1}{2(n-1)}[n+1+(n^2+10n-7)^{\frac{1}{2}}]$, $n = 2,3$,

$1 < p < \infty$, $n = 1$.

Then $T < \infty$ and u cannot exist for all time.

B) Let $n = 3$, $p = 1+\sqrt{2}$ and (ii) hold. Then $T < \infty$ and u cannot exist for all time provided the integrals of f and g in (ii) are sufficiently large.

7 Decay results

The previous §§ have considered nonexistence by means suggestive of blow-up. For those cases the nonlinearities were typically superlinear, eg. u^2, u^3, e^u, etc. It is natural to wonder what happens when f(u) is a sublinear function, i.e. $f = u^\alpha$, $0<\alpha<1$, $u > 0$. To make some remarks on this question we look at the work of Knops & Straughan [99]. In addition, this § discusses the behaviour of a sublinear heat equation on an unbounded spatial region and a heat equation with a sublinear conductivity.

Let us again consider (2.1) with f and g as in that §, although we shall now require that the nonlinearities satisfy the sublinear restriction

$$\langle u,f\rangle + 2(2\alpha-1)G(u) \leq \mu F^\gamma, \tag{7.1}$$

where now

$$F(t) = \|u(t)\|^2 , \tag{7.2}$$

for real numbers α, μ, γ with $\alpha>1$, $\mu>0$, $0<\gamma<1$. (For example, if $u>0$, $f=u^\beta$, $0<\beta<1$, then

$$G(u) = \frac{1}{(1+\beta)}\int_\Omega u^{1+\beta}dx$$

and (7.1) becomes,

$$\langle u,f\rangle+2(2\alpha-1)G(u) \leq \left(\frac{\beta+4\alpha-1}{1+\beta}\right)m^{(1-\beta)/2}F^{(1+\beta)/2},$$

where m = volume of Ω.)

Note that the energy equation is again,

$$E(t) \equiv \tfrac{1}{2}\,||u_t||^2 + \tfrac{1}{2}\,||\text{grad}u||^2 - G(u) = E(0). \tag{7.3}$$

Underlying the approach to decay is a variant of the method of §2 which shows $(F^{\alpha})'' \leq 0$; an idea due to B.D.Sleeman. Therefore, we wish to consider $FF'' + (\alpha-1)(F')^2$ and so compute

$$F'' = 2\,||u_t||^2 - 2\,||\text{grad}u||^2 + 2\langle u,f\rangle,$$

and it follows that

$$FF''+(\alpha-1)(F')^2 = -4(\alpha-1)\{\,||u||^2\,||u_t||^2 - \langle u,u_t\rangle^2\}$$

$$+2(2\alpha-1)F\,||u_t||^2 - 2F\,||\text{grad}u||^2 + 2F\langle u,f\rangle.$$

The first term is $\leqq 0$ by Cauchy-Schwarz and we substitute for the $||u_t||^2$ term from (7.3) to find

$$FF''+(\alpha-1)(F')^2 \leqq -4\alpha F\,||\text{grad}u||^2 + 4(2\alpha-1)FE(0)$$

$$+2F[\langle u,f\rangle + 2(2\alpha-1)G(u)]$$

$$\leqq 4(2\alpha-1)FE(0) + 2\mu F^{1+\gamma}. \tag{7.4}$$

To proceed further it seems necessary to assume

$$E(0) \leqq 0. \tag{7.5}$$

Assuming this we discard the E(0) term and then multiply (7.4) by $\alpha F^{\alpha-2}$; the result is

$$(F^{\alpha})'' \leqq 2\mu\alpha F^{\alpha-\varepsilon}, \tag{7.6}$$

where $\varepsilon = 1-\gamma$.

It is now necessary to further restrict the initial data to satisfy

$$F'(0) < -\frac{\xi}{\alpha} F^{1-\varepsilon/2}(0), \tag{7.7}$$

where $\xi = 2\alpha[\mu/(2\alpha-\varepsilon)]^{\frac{1}{2}}$.

Let now $H = F^{\alpha}$. If $F \to 0$ for some finite t we are finished. So suppose $F > 0$, $t > 0$. By assumption $F'(0)<0$, and so either $F'<0$ on $[0,\infty)$ or there is an $\eta<\infty$ such that $F'(\eta) = 0$. Suppose the second alternative holds. Then, on $[0,\eta)$ multiply (7.6) by F' and integrate to find

$$P(t) \geq P(0) > 0, \quad 0 \leq t < \eta, \tag{7.8}$$

where

$$P(t) = (H'(t)-\xi H(t)^{1-\varepsilon/2\alpha})(H'(t)+\xi H(t)^{1-\varepsilon/2\alpha}). \tag{7.9}$$

By continuity, therefore, $P(\eta) \neq 0$ and since $F(\eta) > 0$, $F'(\eta) < 0$, contradiction. Therefore (7.8) holds for all t for which the solution exists. Under these conditions neither factor of $P(t)$ may change sign and so

$$H' + \xi H^{1-\varepsilon/2\alpha} \leq 0. \tag{7.10}$$

Integrating this we are led to

$$F(t) \leq \left(F^{(1-\gamma)/2}(0) - t(1-\gamma)\left[\frac{\mu}{2\alpha+\gamma-1}\right]^{\frac{1}{2}}\right)^{2/(1-\gamma)}. \tag{7.11}$$

Hence, from (7.11) either $F \to 0$ for some time before or equal to $t^{\#} = \frac{[F(0)]^{(1-\gamma)/2}}{(1-\gamma)}\left(\frac{2\alpha+\gamma-1}{\mu}\right)^{\frac{1}{2}}$, or u ceases to exist before it reaches zero.

The above results have all been concerned with the case where Ω is bounded. We turn now to a decay result when Ω is of infinite extent.

Heat conduction in an infinite body (Knops & Straughan [99])

Payne [146] has shown that a solution to the heat equation with a sublinear heat sink must decay to zero in a finite time if the region is bounded (or if the region is unbounded and decay restrictions are placed on the solution as the spatial variables tend to infinity). We here establish a complementary result to that of Payne by showing his result remains true if Ω is the complement of a bounded set in R^3 without prescribing artificial decay restrictions at infinity. The method used to prove this result relies on a combination of Payne's [146] proof and weighted L^p estimates similar to those of Galdi & Rionero [43].

Let now Ω_o be a bounded domain in R^3 with boundary Σ. Define Ω by

$$\Omega = R^3 - \bar{\Omega}_o .$$

Let $B(x,\varepsilon)$ denote the open ball, centre x, radius ε, and let $S(x,\varepsilon)$ denote the closed surface surrounding $B(x,\varepsilon)$. Suppose R_o is the first real number such that $B(0,R_o) \supseteq \Omega_o$. Then for $R \geq R_o$, $(R^2 = x_i x_i)$, define

$$\Omega_R = B(0,R) - \bar{\Omega}_o , \tag{7.12}$$

and

$$\Gamma_R = S(0,R). \tag{7.13}$$

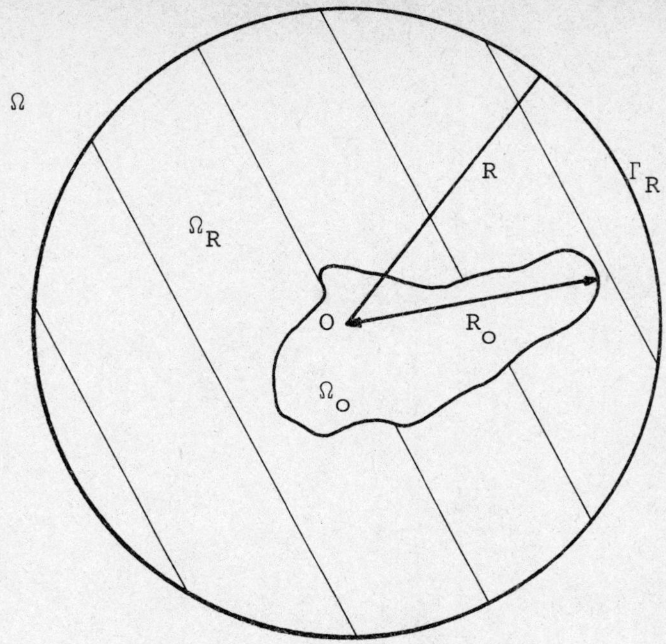

Figure 7.1

We shall consider the problem

$$\begin{aligned} \frac{\partial u}{\partial t} &= \Delta u - f(u), && \text{in } \Omega\times(0,T), \\ u &= 0, && \text{on } \Sigma\times[0,T), \\ u(x,0) &= u_o(x), \end{aligned} \tag{7.14}$$

where, as in Payne [146], the point function f(u) satisfies the inequality

$$uf(u) \geqq k|u|^{2-\varepsilon}, \quad 0 < \varepsilon < 1, \tag{7.15}$$

for some positive constant k.

Since we are assuming no decay restrictions on u at infinity it is necessary to impose the following growth conditions

$$|u| \leqq M, \qquad \left|\frac{\partial u}{\partial r}\right| \leqq Mr^{\theta}, \quad \text{in } \Omega, \tag{7.16}$$

for <u>any</u> $\theta > 0$ and for all $r > 0$.

Define the function g by

$$g(r,t) = \exp(-\alpha[t+t_o]r), \tag{7.17}$$

for positive constants t_o, α to be chosen, and further define the weighted L^p integrals

$$\Phi_{p,R}(t) = \int_{\Omega_R} g(r,t)u^{2p}dx, \tag{7.18}$$

for a positive integer p and

$$\Phi_p(t) = \int_{\Omega} gu^{2p}dx. \tag{7.19}$$

Now

$$\begin{aligned} \frac{d}{dt}\Phi_{p,R} &= \int_{\Omega_R} g_t u^{2p}dx + 2p\int_{\Omega_R} gu^{2p-1}u_t dx, \\ &= -\alpha\int_{\Omega_R} rgu^{2p}dx - 2p\int_{\Omega_R} gu^{2p-1}f(u)dx \\ &\qquad + 2p\int_{\Omega_R} gu^{2p-1}\Delta u dx. \end{aligned} \tag{7.20}$$

Integration by parts on the last term in (7.20) gives

$$\begin{aligned} 2p\int_{\Omega_R} gu^{2p-1}\Delta u dx &= -2p(2p-1)\int_{\Omega_R} gu^{2p-2}|\text{grad}u|^2dx \\ &+\int_{\Omega_R} \Delta g.u^{2p}dx - \int_{\Gamma_R} \frac{\partial g}{\partial r}u^{2p}dS + 2p\int_{\Gamma_R} gu^{2p-1}\frac{\partial u}{\partial r}dS. \end{aligned}$$

We put this into (7.20) and then let $R \to \infty$ to find

$$\frac{d}{dt}\Phi_p = -\alpha\int_{\Omega} rgu^{2p}dx - 2p\int_{\Omega} gu^{2p-1}f(u)dx$$

$$-2p(2p-1)\int_\Omega gu^{2p-2}|\mathrm{grad}u|^2dx + \int_\Omega u^{2p}\Delta g dx. \tag{7.21}$$

Further,

$$g_{,i} = -\alpha(t+t_o)g.\frac{x_i}{r},$$

$$\Delta g = \alpha^2(t+t_o)^2 g,$$

and so we use this in (7.21) and discard the $|\mathrm{grad}u|^2$ term to see that

$$\frac{d}{dt}\Phi_p \leq -2p\int_\Omega gu^{2p-1}f(u)dx+\alpha\int_\Omega gu^{2p}[\alpha(t+t_o)^2-r]dx. \tag{7.22}$$

Now, fix $t_o > 0$ and then select α such that

$$\alpha < \frac{R_o(k\varepsilon)^2}{(Q^\varepsilon+t_o k\varepsilon)^2}, \tag{7.23}$$

where

$$Q = \lim_{p\to\infty}[\Phi_p(0)]^{1/2p}, \tag{7.24}$$

and so for $t < Q^\varepsilon/k\varepsilon$ we may discard the last term in (7.22). We do this and then employ (7.15), to obtain

$$\Phi_p' \leq -2pk\int_\Omega g|u|^{2p-\varepsilon}dx,$$

$$\leq \frac{-2pk}{[u^*(t)]^\varepsilon}\Phi_p ,$$

where, as in Payne [146],

$$u^*(t) = \sup_{x\in\Omega}|u(x,t)|. \tag{7.25}$$

This inequality integrates to

$$\Phi_p(t) \leq \Phi_p(0)\exp\left(-2pk\int_0^t \frac{d\eta}{[u^*(\eta)]^\varepsilon}\right). \tag{7.26}$$

Next, bound $\Phi_p(t)$ below by $\Phi_{p,R}(t)$ for some $R \geqq R_o$, and we may establish

$$\left(\int_{\Omega_R} u^{2p}dx\right)^{1/2p} \leqq \exp\left(\frac{\alpha(t+t_o)R}{2p} - k\int_0^t \frac{d\eta}{[u^*(\eta)]^{\varepsilon}}\right)[\Phi_p(0)]^{1/2p}.$$

Letting $p \to \infty$ now shows that

$$\sup_{x \in \Omega_R} |u(x,t)| \leqq Q\exp\left(-k\int_0^t \frac{d\eta}{[u^*(\eta)]^{\varepsilon}}\right).$$

However, R is arbitrary and so

$$u^*(t) \leqq Q\exp\left(-k\int_0^t \frac{d\eta}{[u^*(\eta)]^{\varepsilon}}\right). \tag{7.27}$$

Now, raise both sides of (7.27) to the power ε and integrate to find that

$$\frac{t}{Q^{\varepsilon}} \leqq \frac{1}{k\varepsilon}\left[1-\exp\left(-k\varepsilon\int_0^t \frac{d\eta}{[u^*(\eta)]^{\varepsilon}}\right)\right] < \frac{1}{k\varepsilon}\,. \tag{7.28}$$

Therefore, either u ceases to exist before $t_1 = Q^{\varepsilon}/k\varepsilon$ or u becomes zero for some $t \leqq t_1$. Observe that since our choice of α in (7.23) is compatible with this value of t_1 we have shown that a non-trivial bounded solution to (7.14) with radial derivatives which grow no faster than a polynomial cannot exist for all time.

Another entirely different use of the weighted energy method to derive a decay result is due to Morawetz [125]. She shows that if u is a solution to the wave equation $u_{tt} = \Delta u$ in the exterior of a smooth, three-dimensional body Ω_o with $u = 0$ on the boundary Σ and the initial data for u and u_t vanish on Σ and have compact support, then $|u(\underset{\sim}{x},t)|$, for any $\underset{\sim}{x} \notin \bar{\Omega}_o$, decays at least as fast as $t^{-\frac{1}{2}}$ if Ω_o is star-shaped. The proof of this result involves estimating $|u|$ by a kind of Sobolev inequality and an interesting weighted energy identity which results by multiplying the wave equation by $x^i u_{,i} + tu_t + u$.

A fast diffusion problem arising in plasma physics.

To complete this § we consider another type of sublinear heat equation. A series of interesting papers by Berryman and his co-workers [7-10,23] derived stability and decay results for a nonlinear diffusion equation which arises in a study of cross-field plasma transport in toroidal multipole magnetic confinement configurations. In its simplest form the boundary-initial value problem they studied is

$$\frac{\partial n}{\partial t} = \frac{\partial^2}{\partial x^2}(n^{\delta+1}), \quad 0 < x < 1, \ t > 0,$$

$$n = 0, \quad x = 0,1, \quad \forall t \geqq 0, \tag{7.29}$$

$$n(x,0) = n_o(x),$$

where δ is a constant with either $\delta>0$, slow diffusion, or $\delta<0$, fast diffusion and $n(x,t)$ is the plasma density profile. Physically, therefore, $n > 0$ and so attention is restricted to positive solutions.

When $\delta>0$ (7.29) has also been used to model flow in a porous medium and Levine & Payne [116] have shown by concavity arguments that solutions to the backward in time problem cannot exist for all time. For the forward in time problem, however, solutions decay in time and Berryman [8] remarks that the asymptotic time dependence is a negative power law. It is interesting to note that an estimate of the decay is easily obtained using ideas similar to those of §4. To see this define F(t) by

$$F(t) = \int_0^1 \phi n dx, \tag{7.30}$$

where $\phi(x)$ is the first eigenfunction in the membrane problem, i.e.

$$\phi'' + \lambda_1 \phi = 0, \quad 0 < x < 1, \tag{7.31}$$

with $\phi(0) = \phi(1) = 0$. For positive n, F is clearly positive.

Differentiating F, using (7.29) and integrating by parts we find

$$F' = \int_0^1 \phi'' n^{1+\delta} dx,$$

$$= -\lambda_1 \int_0^1 \phi n^{1+\delta} dx,$$

by (7.31). From Jensen's inequality

$$\int_0^1 \phi n^{1+\delta} dx \geqq F^{1+\delta}, \tag{7.32}$$

and so

$$F' + \lambda_1 F^{1+\delta} \leqq 0.$$

This easily integrates to yield

$$F(t) \leqq \frac{F(0)}{(1+\lambda_1 t\delta[F(0)]^{\delta})^{1/\delta}}, \tag{7.33}$$

from which the required decay follows.

The above method works with $(n^{1+\delta})_{xx}$ replaced by any convex nonlinearity f_{xx}, provided $f(0) = 0$, as all we require is (7.32) to be valid.

The situation in which $-1 < \delta < 0$ occupies most of the attention in [7-10,23]. It can be shown in this case that there is a finite time t^* at which the solution first vanishes identically and one of the objects in [7-10,23] was to estimate t^*. We now describe results of Berryman & Holland [9] which derive an upper and a lower bound for t^*. Consistent with earlier §§ we concentrate on differential inequality arguments and do not describe the more involved eigenfunction expansion techniques given in [7-10,23] as these are developed primarily for the special problem (7.29). We should mention, however, that in [8] an asymptotic series is derived for the exact value of t^*.

For simplicity suppose $\delta = -\frac{1}{2}$ and suppose there exists a first time t^* for which $n(x,t^*) \equiv 0$. Then introduce $u(x,t) = n^{\frac{1}{2}}(x,t)$. Equation (7.29) becomes

$$2uu_t = u_{xx}, \quad 0 < x < 1,\ t > 0, \tag{7.34}$$

with homogeneous boundary conditions and $u(x,0)$ given.

Theorem 7.1 (Berryman & Holland [9]).

$$2\int_0^1 u^3(x,0)\,dx \Big/ \int_0^1 u_x^2(x,0)\,dx \leq t^* \leq$$

$$2\left(\frac{3\pi}{2^7}\right)^{2/3}\int_0^1 u^3(x,0)\,dx. \tag{7.35}$$

Proof. Let all integrals in this proof be over (0,1) unless otherwise indicated and define $Q(t)$, $R(t)$ by

$$Q(t) = \int u^3 dx, \qquad R(t) = \int u_x^2 dx.$$

Then, with the aid of (7.34),

$$Q' = -3R/2, \quad R' = -\int \frac{u_{xx}^2}{u} dx. \tag{7.36}$$

The derivation of the upper bound for t^* begins with

$$u(x,t) = \int_0^x u_x dx \leq x^{\frac{1}{2}} [\int_0^x u_x^2 dx]^{\frac{1}{2}},$$

so

$$u^2(x,t) \leq x \int_0^x u_x^2 dx. \tag{7.37}$$

Similarly, representing u as an integral from x to 1, we find

$$u^2(x,t) \leq (1-x) \int_x^1 u_x^2 dx. \tag{7.38}$$

Combining (7.37) and (7.38) leads to

$$u^2(x,t) \leq x(1-x) \int u_x^2 dx. \tag{7.39}$$

From this inequality easily follows

$$Q^{2/3} \leq R(t)[\beta(5/2,5/2)]^{2/3}, \tag{7.40}$$

β denoting the β-function. Substituting $R = -2Q'/3$ in this inequality and integrating we find

$$-\frac{d}{dt} Q^{1/3} \geq \tfrac{1}{2} [\beta(5/2,5/2)]^{-2/3}.$$

This expression is integrated from 0 to t^* and the right hand side of (7.35) follows, after $\beta(5/2,5/2)$ is calculated, since $Q(t^*) = 0$.

To derive the lower bound proceed as follows.

$$R^2 = [\int u_x^2 dx]^2 = [\int u u_{xx} dx]^2 \leq \int u^3 dx \int \frac{u_{xx}^2}{u} dx \ ,$$

by Cauchy-Schwarz. Employ both of (7.36) in the last inequality to obtain

$$QR' \leq \frac{2}{3}RQ'.$$

After separating variables this inequality may be integrated to give

$$R \leq R(0)Q^{2/3}[Q(0)]^{-2/3}.$$

However, $R = -2Q'/3$ and the above inequality therefore yields

$$\frac{d}{dt}Q^{1/3} \geq -\tfrac{1}{2}R(0)/Q(0)^{2/3}.$$

An integration of this from 0 to t^* shows that

$$t^* \geq 2Q(0)/R(0),$$

which is the lower bound in (7.35).

To complete our study of nonexistence we return to the topic of blow-up and analyse two parabolic problems from fluid mechanics.

8 Nonexistence theorems for problems in nonlinear fluid theories

A careful study of thermodynamics, stability and instability for third grade fluids has recently been published by Fosdick & Rajagopal [30], this having been preceded by a similar study for second grade fluids by Dunn & Fosdick [24]. We here describe a result of Fosdick & Straughan [31] which points out an apparent important difference between fluids of third grade and those of second grade.

The equations of (isothermal) motion for a fluid of second or third grade are (see [24,30]):

$$\rho\dot{v}_i = \rho b_i + T_{ji,j}\,, \qquad v_{i,i} = 0, \tag{8.1}$$

where $\underset{\sim}{v}$, ρ, $\underset{\sim}{b}$, $\underset{\sim}{T}$ are respectively, the fluid velocity, (constant) density, external body force and stress tensor. A superposed dot denotes the material derivative and (8.1) are to be defined on a bounded region, $\Omega(t)$, of three-space, for all time.

Let $\underset{\sim}{L}$ be the velocity gradient tensor and then, in terms of the first two Rivlin-Ericksen tensors

$$\underset{\sim}{A}_1 = \underset{\sim}{L} + \underset{\sim}{L}^T, \tag{8.2}$$

$$\underset{\sim}{A}_2 = \dot{\underset{\sim}{A}}_1 + \underset{\sim}{A}_1\underset{\sim}{L} + \underset{\sim}{L}^T\underset{\sim}{A}_1 \tag{8.3}$$

the constitutive equation for a fluid of second grade is (see [24]),

$$\mathbf{T} = -p\mathbf{I} + \mu\mathbf{A}_1 + \alpha_1\mathbf{A}_2 + \alpha_2\mathbf{A}_1^2 , \qquad (8.4)$$

where the coefficients in (8.4) are restricted by the Clausius-Duhem inequality to satisfy

$$\mu \geqq 0, \qquad \alpha_1+\alpha_2 = 0. \qquad (8.5)$$

By employing the Clausius-Duhem inequality <u>and</u> demanding that the free energy be a minimum in equilibrium, Fosdick & Rajagopal [30] have shown that the corresponding constitutive equation for a fluid of third grade is

$$\mathbf{T} = -p\mathbf{I} + \mu\mathbf{A}_1 + \alpha_1\mathbf{A}_2 + \alpha_2\mathbf{A}_1^2 + \beta(\mathrm{tr}\mathbf{A}_1^2)\mathbf{A}_1, \qquad (8.6)$$

where now

$$\mu \geqq 0, \quad \beta \geqq 0, \quad -(24\mu\beta)^{\frac{1}{2}} \leqq \alpha_1+\alpha_2 \leqq (24\mu\beta)^{\frac{1}{2}}, \qquad (8.7)$$

$$\alpha_1 \geqq 0. \qquad (8.8)$$

It is important to note that the Clausius-Duhem inequality alone yields no information on the sign of α_1 in (8.4), or on the sign of either α_1 or α_2 in (8.6) (although the form (8.6) emerges only after compatibility with thermodynamics is demanded and the free energy is a minimum in equilibrium, for which case $\alpha_1 \geqq 0$, see [30]). Dunn & Fosdick [24] and Fosdick & Rajagopal [30] have shown, however, that $\alpha_1 < 0$ can lead to an exponential type of instability. Since some viscometric measurements would indicate $\alpha_1 < 0$ such findings are obviously important to the study of viscoelastic fluids.

Suppose a fluid of grade three is undergoing a mechanically isolated cannister flow; that is $\Omega(t)$ $(\equiv\Omega)$ is fixed for all $t \geqq 0$, $\mathbf{v} = \mathbf{0}$ on the boundary Γ and

$$\int_\Omega \rho b_i^\# v_i dx = 0, \tag{8.9}$$

where $b_i^\#$ is the part of the body force not derivable from a potential. Suppose also that the strict form of inequalities (8.7) hold and

$$\alpha_1 < 0; \tag{8.10}$$

thus, the restrictions imposed by the Clausius-Duhem inequality are met, but the free energy is not a minimum in equilibrium. These conditions correspond to those of §6 in the paper of Fosdick & Rajagopal [30].

In [30] (lemma 3) it is shown that (8.7) are equivalent to

$$\mu|\underset{\sim}{A}|^2 + (\alpha_1 + \alpha_2)\mathrm{tr}\underset{\sim}{A}^3 + \beta|\underset{\sim}{A}|^4 \geq 0, \tag{8.11}$$

where $\underset{\sim}{A}$ is any non-zero, symmetric, second order tensor with zero trace. Since we are supposing (8.7) hold strictly, then there exists $\varepsilon(>0)$ such that both (8.7) hold strictly and consequently, (8.11) holds with β replaced by $\beta^\# \equiv \beta-\varepsilon$.

Define N(t), for $t \geq 0$, by

$$N(t) = \frac{|\alpha_1|}{2\rho}\int_\Omega |\underset{\sim}{A}_1|^2 dx - \int_\Omega |\underset{\sim}{v}|^2 dx. \tag{8.12}$$

Fosdick & Rajagopal [30], theorem 7, have shown that

$$N(t) \geq N(0)\exp\left(\frac{t(24\mu\beta-[\alpha_1+\alpha_2]^2)}{12\beta|\alpha_1|}\right)$$

and so if $N(0)>0$, then N(t) grows exponentially with time. We show that when $N(0)>0$ the solution to (8.1) cannot exist for all time. (With the aid of the Poincaré inequality, it is always possible to arrange that $N(0)>0$, by choosing a small enough container.)

Using (8.1), (8.6), (8.9), (8.10) and the boundary condition $v_i \equiv 0$, one may show ([30], eq. (6.4)),

$$\dot{N}(t) = \frac{\mu}{\rho}\int_\Omega |\underset{\sim}{A}_1|^2 dx + \frac{(\alpha_1+\alpha_2)}{\rho}\int_\Omega \operatorname{tr}\underset{\sim}{A}_1^3 dx + \frac{\beta}{\rho}\int_\Omega |\underset{\sim}{A}_1|^4 dx. \tag{8.13}$$

However, $\beta = \beta^{\#}+\varepsilon$ and (8.11) holds with β replaced by $\beta^{\#}$ and so from (8.13) and (8.11) we deduce that

$$\dot{N}(t) \geqq \frac{\varepsilon}{\rho}\int_\Omega |\underset{\sim}{A}_1|^4 dx. \tag{8.14}$$

From the Cauchy-Schwarz inequality

$$\int_\Omega |\underset{\sim}{A}_1|^4 dx \geqq \frac{1}{m}\left(\int_\Omega |\underset{\sim}{A}_1|^2 dx\right)^2,$$

where m = volume of Ω and hence

$$\dot{N}(t) \geqq \frac{\varepsilon}{\rho m}(1-\xi)\left(\int_\Omega |\underset{\sim}{A}_1|^2 dx\right)^2 + \frac{\varepsilon\xi}{\rho m}\left(\int_\Omega |\underset{\sim}{A}_1|^2 dx\right)^2, \tag{8.15}$$

where we shall select ξ to satisfy

$$\xi = \frac{\rho^2}{\rho^2+\alpha_1^2\lambda_1^2}, \tag{8.16}$$

where λ_1 is the constant in Poincaré's inequality which for this case is

$$\int_\Omega |\underset{\sim}{A}_1|^2 dx \geqq 2\lambda_1\int_\Omega |\underset{\sim}{v}|^2 dx.$$

This is used in (8.15) with the result

$$\dot{N}(t) \geqq \phi\left[\frac{|\alpha_1|^2}{4\rho^2}\left(\int_\Omega |\underset{\sim}{A}_1|^2 dx\right)^2 + \left(\int_\Omega |\underset{\sim}{v}|^2 dx\right)^2\right], \tag{8.17}$$

where ϕ is the positive constant

$$\phi = \frac{4\varepsilon\lambda_1^2\rho}{m(\rho^2+\lambda_1^2\alpha_1^2)}. \tag{8.18}$$

Completing the square in (8.17) we find

$$\dot{N}(t) \geqq \phi N^2(t). \tag{8.19}$$

Suppose now $N(0)>0$, so that $N(t)>0$ in some positive neighbourhood of $t = 0$. Then, in this neighbourhood we may rewrite (8.19) as

$$-\frac{d}{dt}[1/N(t)] \geqq \phi,$$

which integrates to

$$N(t) \geqq \frac{N(0)}{1-N(0)\phi t}. \tag{8.20}$$

But then by the same reasoning the neighbourhood in which (8.20) holds may be enlarged and accordingly we see that N cannot exist beyond $T = [\phi(N(0))]^{-1}$, and, from the definition of N, neither can $\int_\Omega |\underset{\sim}{A}_1|^2 dx$, and so we conclude that the solution v_i cannot exist for all time

Fosdick & Rajagopal [30] assert that $\alpha_1 < 0$ being sufficient to guarantee "exponential instability" gives rise to a fundamentally unstable, and therefore unacceptable, theory of fluid behaviour; our analysis offers additional weight to their assertion. It is interesting to observe that the behaviour of a mechanically isolated cannister flow with $\alpha_1 < 0$ as predicted by (8.20) is very different to the corresponding situation when the fluid of grade two is employed; for, Dunn & Fosdick [24], theorem 18, show that N(t) and $\int_\Omega |\underset{\sim}{A}_1|^2 dx$ are bounded above by an exponential function of time for the second grade fluid theory, whereas by (8.20) the theory for a fluid of third grade shows that if N(t) or $\int_\Omega |\underset{\sim}{A}_1|^2 dx$ exist long enough they become infinite in a finite time.

It is also interesting to note that the larger is ε, i.e. the larger the viscosity coefficient β, the smaller is T. This would imply that for a fluid of third grade with $\alpha_1<0$, the larger β holding everything else fixed, the more rapidly break-down occurs. A similar anomalous growth behaviour was found regarding μ by Fosdick & Rajagopal [30].

Another worthwhile observation in connection with the fluid of grade three is that if $\alpha_1>0$, the other conditions of this § being the same, then (8.13) and (8.14) lead to the conclusion that $\int_\Omega |\underset{\sim}{A}_1|^4 dx \in L^1(0,\infty)$ and this implies decay as $t \to \infty$, except perhaps for a countable number of intervals of vanishingly small measure. However, as $W_0^{1,4}(\Omega) \subset C^0(\Omega^-)$, see Gilbarg & Trudinger [52] p.148, one should be able to predict from the third grade fluid model decay in time of $|\underset{\sim}{v}|$, i.e. pointwise decay. In this sense, the viscoelastic nature of the model is leading to a smoother solution than that for the Navier-Stokes equations.

The final topic covered in this § is the behaviour of the solution to a set of equations proposed by Ladyzhenskaya as possible alternatives to the usual equations of viscous flow (see [173]).

As alternatives to the Navier-Stokes equations, Ladyzhenskaya has proposed the following sets of equations,

$$\dot{u}_i - (\nu_0+\nu_1\int_\Omega u_{m,n}u_{m,n}dx)\Delta u_i = -p_{,i} + f_i, \qquad (8.21)$$

$$\dot{u}_i - ([\nu_0+\nu_1(u_{m,n}u_{m,n})]u_{i,j})_{,j} = -p_{,i} + f_i, \qquad (8.22)$$

$$\dot{u}_i + (\mathrm{curl}([\nu_0+\nu_1|\mathrm{curl}\,u|^2]\mathrm{curl}\,u))_i = -p_{,i} + f_i, \qquad (8.23)$$

where $\underset{\sim}{u}$ is again a solenoidal velocity field and apart from standard notation ν_1 is a second viscosity coefficient.

We show that a solution to the boundary-initial value problem for (8.21) does not exist globally backward in time. (The analysis for the other models is very similar and is covered in [173].)

Suppose, therefore, that (8.21) are defined on $\Omega\times(-T,0)$, $\Omega\subset R^3$ bounded, with $u_i \equiv 0$ on Γ, $f_i \equiv 0$ and

$$u_i(x,0) = v_i(x). \qquad (8.24)$$

Theorem 8.1. Let $u_i \in C^2(\Omega^-\times(-T,0])$, $i = 1,2,3$, $p\in C^1(\Omega^-\times(-T,0])$, be a solution to (8.21) with $\underset{\sim}{u} \equiv \underset{\sim}{0}$ on Γ, (8.24) as initial data and $v_i(x) \not\equiv 0$. Then $u_i(x,t)$ cannot exist for all negative time. Furthermore, an upper bound for the length of the possible interval of existence, $(-T,0]$, is given by T^*, where

$$T^* = \frac{1}{(2\nu_0\lambda_1)}\log[1+\nu_0/\nu_1\lambda_1\,\|v\|^2\,],$$

where λ_1 is the Poincaré constant and $\|v\|^2 = \int_\Omega v_i v_i dx$.

Proof. Form the inner product of (8.21) with u_i,

$$\int_\Omega u_i\dot{u}_i dx+\nu_0\int_\Omega u_{i,j}u_{i,j}dx+\nu_1[\int_\Omega u_{i,j}u_{i,j}dx]^2 = 0, \qquad (8.25)$$

as can be seen by using the divergence theorem, boundary

conditions and the fact that

$$\int_{\Omega} p_{,i} u_i dx = 0.$$

Further, since $u_i = 0$ on Γ, we may rewrite (8.25) as

$$\frac{d}{d(-t)} \|u(-t)\|^2 + 2\nu_0 \|\text{grad}u(-t)\|^2 + \nu_1 \|\text{grad}u(-t)\|^4 = 0. \tag{8.26}$$

We now introduce the function $w_i(x,t)$ by,

$$w_i(x,t) = u_i(x,-t), \quad 0 \leqq t < T. \tag{8.27}$$

In terms of w, (8.26) is

$$\frac{d}{dt} \|w(t)\|^2 = 2\nu_0 D(w) + 2\nu_1 [D(w)]^2, \tag{8.28}$$

where $D(w) = \int_{\Omega} w_{i,j} w_{i,j} dx$.

From this point we could use the same type of argument to establish nonexistence as we used for the third grade fluid. However, we shall instead employ a variation of the concavity method.

Define F(t) by

$$F(t) = \int_0^t \|w\|^2 \, d\eta. \tag{8.29}$$

Then, (8.28) is

$$F'' = 2\nu_0 D(w) + 2\nu_1 [D(w)]^2. \tag{8.30}$$

Poincaré's inequality is $D(w) \geqq \lambda_1 \|w\|^2$ and used in (8.30) yields

$$F''(t) - 2\nu_1 \lambda_1^2 (F'(t))^2 \geqq 2\nu_0 \lambda_1 F'(t). \tag{8.31}$$

Upon multiplying (8.31) by $-2\nu_1\lambda_1^2 \exp[-2\nu_1\lambda_1^2 F(t)]$ we deduce that

$$\frac{d^2}{dt^2}[\exp(-2\nu_1\lambda_1^2F(t))] \leqq 2\nu_0\lambda_1\frac{d}{dt}[\exp(-2\nu_1\lambda_1^2F(t))]. \tag{8.32}$$

Integrating and rearranging, we are led to

$$F(t) \geqq \frac{1}{2\nu_1\lambda_1^2}\log\left(\frac{\nu_0}{[\nu_0+\nu_1\lambda_1F'(0)-\nu_1\lambda_1F'(0)\exp(2\nu_0\lambda_1t)]}\right). \tag{8.33}$$

Clearly, therefore, $F(t)$ cannot remain finite for all time and the theorem follows on noting that $F'(0) = \|v\|^2$.

Although the next § is not concerned with nonexistence it studies the related topic of instability and, in particular, derives growth estimates for a linear differential equation. The exponential estimates to be found are typical of linear equations as opposed to the finite time blow-up behaviour of nonlinear ones.

9 Growth estimates appropriate to the equation $Au'' + Bu' + Cu = 0$

In this § we shall show how to apply the weighted energy method to a linear ordinary differential equation in Hilbert space with symmetric operator coefficients. The results described here were obtained by the writer in [176].

Let L(u) denote the operator

$$L(u) \equiv A\frac{d^2u}{dt^2} + B\frac{du}{dt} + Cu, \qquad (9.1)$$

where A, B and C are symmetric linear operators defined on a dense subset, D, of a real Hilbert space H. The norm and inner product on H are denoted by $\|.\|$ and $< \, , >$, respectively. Our object is to make a systematic study of the growth properties of solutions to the equation

$$L(u) = 0, \quad t>0. \qquad (9.2)$$

The behaviour of many physical situations is described by (9.2). In [176] it is shown how the motion of a mixture of two anisotropic, linear elastic solids under isothermal deformations and the motion of a director rod are covered by (9.2); it is also indicated how (9.2) is pertinent to Bénard convection and rotating magnetic stars. Furthermore, when $B \equiv 0$, linear elastodynamics and the many ramifications of this theory, eg. micropolar, multipolar and diatomic elasticity, are capable of description. Growth estimates for the situation $B \equiv 0$ were obtained via logarithmic convexity by Knops & Payne [91] and then with a weighted energy approach similar to that used here by Levine [114] and Hills & Knops [66].

Let A be positive-definite, B be positive and C merely symmetric, i.e.

$$\langle Ax,x\rangle > 0, \quad \forall x\in D,\ x \neq 0,$$

$$\langle Bx,x\rangle \geqq 0, \quad \forall x\in D.$$

Thus, we are imposing no definiteness whatsoever on C. We assume the existence of a solution, u, to (9.2) with $u\in C^2([0,\infty);D)$, $u'\in C^1([0,\infty);D)$ and $u''\in C([0,\infty);D)$. (u' denotes the strong derivative of u with respect to t.)

The energy, E(t), of (9.2) is defined to be

$$E(t) = \tfrac{1}{2}\langle u',Au'\rangle+\tfrac{1}{2}\langle u,Cu\rangle+\int_0^t \langle u',Bu'\rangle ds, \tag{9.3}$$

and by taking the scalar product of (9.2) with u' and integrating it is easily verified that E is constant, i.e.

$$E(t) = E(0), \quad \text{for each } t>0. \tag{9.4}$$

We consider (9.2) subject to the following Cauchy data,

$$u(0) = u_o, \quad \frac{du}{dt}(0) = v_o.$$

To apply the weighted energy method we introduce an auxilliary function, v, defined by

$$f(t)v(t) = u(t), \tag{9.5}$$

where the weight f is a C^2 real-valued function to be specified. In terms of v (9.2) becomes

$$A(fv'' + 2f'v' + f''v) + B(fv' + f'v) + Cfv = 0. \tag{9.6}$$

The type of result obtained for u is determined by the choice of f(t) and the following lemmas develop two estimates for different functions f. We first define functions F and G by

$$F(t) = \langle u, Au \rangle, \tag{9.7}$$

$$G(t) = F(t) + \int_0^t \langle u, Bu \rangle ds. \tag{9.8}$$

Lemma 9.1. Let u be a solution to (9.2). Then, for a(>0) constant, G(t) satisfies

$$G(t) \geqq e^{2at}F(0) + te^{2at}[F'(0) - 2E(0)a^{-1} - \langle u_o, Bu_o \rangle - 2aF(0)]$$

$$+ [a\langle u_o, Bu_o \rangle + E(0)](e^{2at} - 1)/a^2. \tag{9.9}$$

Proof. Choose $f(t) = e^{at}$ in (9.5) and form the inner product of (9.6) with the quantity $e^{-at}v'$ to derive

$$\langle v', Av'' \rangle + a^2 \langle v, Av' \rangle + \langle v', Bv' \rangle + a\langle v', Bv \rangle + \langle v', Cv \rangle \leqq 0. \tag{9.10}$$

We now integrate (9.10), return to the original variable u and bound below the quantity which arises from the $\langle v', Bv' \rangle$ term by

$$\int_0^t \langle u' - au, B(u' - au) \rangle ds.$$

After a little manipulation and use of the energy identity (9.4) we may obtain

$$2E(0) + 2a^2F(t) - aF'(t) - a\langle u, Bu \rangle + 2a^2 \int_0^t \langle u, Bu \rangle ds \leqq$$

$$\leqq e^{2at}[2E(0) + a\langle u_o, Bu_o \rangle + 2a^2F(0) - aF'(0)] - 2a\langle u_o, Bu_o \rangle. \tag{9.11}$$

Estimate (9.9) follows directly by integrating (9.11).

Lemma 9.2. Let u be a solution to (9.2). Then, the following inequality holds for all $t > 0$:

$$G(t) \geqq [aF'(0)-a\langle u_o,Bu_o\rangle-2a^2E(0)-F(0)](t+a)^4/4a^4$$
$$+[E(0)+\langle u_o,Bu_o\rangle a^{-1}](t+a)^2 + \beta, \qquad (9.12)$$

where $a(>0)$ is a constant, and β is a data term.

Proof. Set now, $f(t) = t+a$ in (9.6), where a is a positive constant to be specified. We form the inner product of (9.6) with $v'(t+a)^{-1}$ and discard some positive terms to arrive at

$$\langle v',Av''\rangle+\langle v',Bv'\rangle+(t+a)^{-1}\langle v',Bv\rangle+\langle v',Cv\rangle \leqq 0.$$

After integrating this inequality and returning to the original variables we use the inequality, for $h \geqq 0$,

$$(t+a)^3\int_0^t h(s)(s+a)^{-3}ds \geqq (t+a)\int_0^t (s+a)^{-1}h(s)ds,$$

on the term which results from the $\langle v',Bv'\rangle$ term and with use of the energy equation (9.4) it is easily shown that

$$F'(t)+\langle u,Bu\rangle \geqq [2E(0)+2\langle u_o,Bu_o\rangle a^{-1}](t+a)+\frac{F(t)}{(t+a)}+\int_0^t \frac{\langle u,Bu\rangle ds}{(s+a)}$$
$$+a^{-4}[aF'(0)-F(0)-2a^2E(0)-a\langle u_o,Bu_o\rangle](t+a)^3. \qquad (9.13)$$

Finally, (9.12) follows from (9.13) by integration.

We now state the main theorem of this § which provides, in a precise sense, a growth estimate for a solution to (9.2). The proof of the theorem is omitted as it is not difficult and follows from estimates (9.9) and (9.12) by choosing the constant a according to the hypotheses holding.

Theorem 9.1.

a. If $E(0)<0$, there exists a constant $T\#$ (≥ 0) such that $G(t)$ is bounded below by an increasing exponential on $[T\#,\infty)$.

b. If $E(0)<0$ and $\langle u_o, Bu_o\rangle = 0$, then $G(t)$ is bounded below by an increasing exponential on, (i) $[0,\infty)$ if $F'(0) \geq 0$; (ii) $[t\#,\infty)$, for a constant $t\#>0$, if $F'(0)<0$.

c. Let $E(0)<0$ and $\langle u_o, Bu_o\rangle > 0$. If $F'(0) \geq -\langle u_o, Bu_o\rangle$, $G(t)$ is bounded below by an increasing exponential on $[0,\infty)$; else there is a $t\#>0$ such that $G(t)$ is bounded below by an increasing exponential on $[t\#,\infty)$.

d. If $E(0)=0$ and $F'(0) > \langle u_o, Bu_o\rangle$, then $G(t)$ is bounded below by an increasing exponential on $[0,\infty)$.

e. When $E(0)>0$, $G(t)$ is bounded below by an increasing exponential on $[0,\infty)$ if $F'(0)\geq 4[E(0)F(0)]^{\frac{1}{2}}+\langle u_o, Bu_o\rangle$.

f. If $E(0)>0$, then $G(t)$ is bounded below by an increasing polynomial of order 4 on $[0,\infty)$, provided

$$F'(0) > \langle u_o, Bu_o\rangle + 2[2E(0)F(0)]^{\frac{1}{2}}. \qquad (\#)$$

g. Let $E(0)>0$, $F(0)>0$, and inequality (#) be an equality. Then, $G(t)$ is bounded below by an increasing quadratic on $[0,\infty)$.

Perhaps the most useful purpose of theorem 9.1 are the implications for instability. For, we see that when $G(t)$ is unbounded on $[c,\infty)$, $c \geq 0$, then $\sup_{[0,t]} \langle u, Bu\rangle + t^{-1}\langle u, Au\rangle$ is also unbounded on $[c,\infty)$ and so the solution to (9.2) is unstable. Of course, the precise nature of the growth estimates depend on the operators A and B and without further knowledge of these we are unable to ascertain whether the unboundedness is through the solution itself or its derivatives.

Although we have considered only the growth of $G(t)$, it is possible to derive information on the behaviour of the "energies" $\frac{1}{2}\langle u',Au'\rangle+\int_0^t\langle u',Bu'\rangle ds$ and $\frac{1}{2}\langle u,Cu\rangle$, and estimates are available when B and C are not symmetric: details are contained in [176].

We complete this § by observing that with the choice $f(t) = e^{-at}$, $a>0$, in (9.5) one may establish an asymptotic stability theorem for solutions to (9.2). This weight function was used for a similar purpose by Duvaut & Lions [25], Rauch [158] and the writer [176]. Rauch's work contains, however, an in-depth study using, in addition, semi-group theory of the decay and overdamping phenomena for the equation

$$u_{tt} + \alpha(x)u_t - \Delta u = 0.$$

10 Energy methods and the Navier-Stokes equations

Throughout the next six §§ Ω denotes either an unbounded or a bounded domain in R^3. As mentioned in §1, if Ω is bounded its boundary is denoted by Γ, whereas if Ω is exterior to a compact set in R^3 then the boundary is denoted by Σ. The surfaces Γ and Σ will always be assumed smooth enough to justify applications of the divergence theorem.

A major part of the work described in the following §§ concerns solutions to the Navier-Stokes equations which govern the behaviour of a linearly viscous, incompressible fluid. In view of the qualitative nature of the analyses to be presented it is sufficient to take the density and kinematic viscosity to be unity and the appropriate equations are then

$$
\begin{aligned}
&v_{i,t} + v_j v_{i,j} = -p_{,i} + f_i + \Delta v_i ,\\
&v_{i,i} = 0,
\end{aligned}
\tag{10.1}
$$

where v_i, p, f_i represent, respectively, the velocity, pressure and body force. For $T(>0)$, possibly $T = \infty$, equations (10.1) are assumed to hold on either $\Omega\times(0,T)$ or $\Omega\times(-T,0)$, depending on whether the initial or final value problem is to be studied. To complete the boundary-initial value problem, $\underset{\sim}{v}$ satisfies the following data,

$$\underset{\sim}{v}(\underset{\sim}{x},0) = \underset{\sim}{g}(\underset{\sim}{x}), \quad \text{in } \Omega, \tag{10.2}$$

$$\underset{\sim}{v}(\underset{\sim}{x},t) = \underset{\sim}{h}(\underset{\sim}{x},t), \quad \text{on } \Gamma \text{ or } \Sigma, \tag{10.3}$$

where $\underset{\sim}{g}$ and $\underset{\sim}{h}$ are preassigned functions with $\underset{\sim}{g}$ at least piecewise C^2 in Ω and $\underset{\sim}{h}$ continuously differentiable in $\underset{\sim}{x}$ and t.

When Ω is bounded the boundary-initial (or final) value problem defined by (10.1)-(10.3) will be denoted by P, whereas when Ω is an exterior region P_E will be used. Although it is possible to derive many of the results given here for suitably defined weak solutions we shall restrict attention to classical solutions; thus we require $\underset{\sim}{v}$ to be twice continuously differentiable in the spatial variables and continuously differentiable in time with grad p continuous in $\underset{\sim}{x}$ and t.

The questions of stability, uniqueness and related properties for solutions to the Navier-Stokes equations can be divided into four sections:

a. The boundary-initial value problem when Ω is bounded.

b. The boundary-final value problem when Ω is bounded.

c. The boundary-initial value problem when Ω is unbounded.

d. The boundary-final value problem when Ω is unbounded.

Case (a) has been widely studied and for this case the weighted energy method is usually not necessary, in the sense that the weight function $g(\underset{\sim}{x},t) \equiv 1$ is sufficient. In particular, the practically important studies of energy stability which give sharp bounds guaranteeing asymptotic stability are beautifully described in depth by Joseph [74]. The conventional energy method ($g \equiv 1$) is very powerful and in some cases leads to a description of stability which exactly complements instability analyses of linear theory; such a case is convection in the Bénard problem, see Joseph [73]. In this article we do not, therefore, pay attention to case (a). Instead, in §§ 11-15, various aspects of cases (b)-(d) are dealt with and we illustrate how different choices of weight function lead to different results and are necessary for the separate cases. Although we deal mostly with the Navier-Stokes equations this is by choice as it seems preferable to illustrate the methods with the simplest fluid theory without combined, competing effects such as non-uniform temperature in the Bénard problem. Nevertheless, §§ 16-18 do involve use of the weight function in other theories, namely, a theory of fluid jet flow, the magnetohydrodynamic theory for the geo-astrophysical Bénard problem and a model for instability of a suspension of micro-organisms.

11 Serrin's backward uniqueness theorem for solutions to the Navier-Stokes equations

In this § we consider solutions to P on $\Omega\times(-T,0)$. In reference [169] Serrin provided an elegant proof of uniqueness of classical solutions backward in time, by implementing the ideas of Lees & Protter [107].

We shall describe Serrin's proof but first it is useful to introduce some further notation different from that of the earlier chapters. For functions a_i, b_i, write

$$|a|^2 = \int_\Omega a_i a_i dx, \qquad \|a\|^2 = \int_\Omega a_{i,j} a_{i,j} dx,$$
$$(a,b) = \int_\Omega a_i b_i dx. \tag{11.1}$$

Define further the linear operator L by

$$Lu = \Delta u - u_t, \tag{11.2}$$

where u is, in general, a vector valued function.

Serrin's proof of backward uniqueness proceeds via the following lemma, where a weight function dependent only on t is introduced. Let $\tau < T$ with $0 < \tau < 1$ be fixed and set $\lambda = t-\tau$.

<u>Lemma 11.1</u>. Let $\phi(x,t)$ be a twice continuously differentiable divergence free vector which is zero on the boundary of the space-time region $\Omega\times(-\tau,0)$. Suppose $\tilde{p}(x,t)$ is an arbitrary continuously differentiable function. Then, for any integer $m > 0$, there holds

$$\int \lambda^{-2m}\{|\phi|^2 + \|\phi\|^2\}\,dt \leqq c\int \lambda^{-2m}|L\phi-\text{grad}\tilde{p}|^2 dt, \qquad (11.3)$$

where $c = 5(\tau+m^{-1})$ and where the integration is over $(-\tau,0)$.

Proof. Define $\psi = \lambda^{-m}\phi$. Then

$$\lambda^{-m}(L\phi-\text{grad}\tilde{p}) = \Delta\psi-\psi_t-m\lambda^{-1}\psi-\lambda^{-m}\text{grad}\tilde{p}.$$

Multiply both sides of the equation by $-\psi_t$, integrate over Ω and use inequality (AG), to arrive at

$$\lambda^{-2m}|L\phi-\text{grad}\tilde{p}|^2 \geqq -2(\psi_t,\Delta\psi)+2m\lambda^{-1}(\psi,\psi_t)+2\lambda^{-m}(\text{grad}\tilde{p},\psi_t) .$$

Since $\psi \equiv 0$ on Γ and $\psi_{i,i} = 0$ it is easily seen that the last term vanishes. Moreover, the first term may similarly be reduced to the form $2(\nabla\psi,\nabla\psi_t)$. Therefore, we find

$$\lambda^{-2m}|L\phi-\text{grad}\tilde{p}|^2 \geqq \frac{d}{dt}\{\|\psi\|^2 + m\lambda^{-1}|\psi|^2\}+m\lambda^{-2}|\psi|^2.$$

An integration over $(-\tau,0)$ leads to

$$\int \lambda^{-2m}|L\phi-\text{grad}\tilde{p}|^2 dt \geqq \int m\lambda^{-2m-2}|\phi|^2 dt. \qquad (11.4)$$

In addition,

$$-\int(\lambda^{-m-1}\phi,\lambda^{-m+1}[L\phi-\text{grad}\tilde{p}])dt = \int \lambda^{-2m}(\|\phi\|^2 + m\lambda^{-1}|\phi|^2)dt ,$$

as can be seen with the aid of the facts that $\phi_{i,i} = 0$ and ϕ is zero on the boundary of $\Omega\times(-\tau,0)$. This expression is now re-arranged and after use of the inequality (AG) we deduce

$$\int \lambda^{-2m}\|\phi\|^2 dt \leqq \tfrac{1}{2}\int \lambda^{-2m-2}|\phi|^2 dt + \tfrac{1}{2}\int \lambda^{-2m+2}|L\phi-\text{grad}\tilde{p}|^2 dt$$
$$+ m\int \lambda^{-2m}|\lambda^{-1}||\phi|^2 dt,$$

$$\leq \tfrac{1}{2}\int\lambda^{-2m-2}|\phi|^2dt+2\tau^2\int\lambda^{-2m}|L\phi-\text{grad}\tilde{p}|^2dt+2\tau m\int\lambda^{-2m-2}|\phi|^2dt, \qquad (11.5)$$

where to obtain the last line observe that

$$|\lambda^{-1}| \leq 2\tau\lambda^{-2}, \quad \lambda^2 \leq 4\tau^2, \qquad (11.6)$$

results which follow since

$$\lambda^2 = (t-\tau)^2 \leq 4\tau^2$$

and so

$$|\lambda| \leq 2\tau.$$

Inequality (11.4) is multiplied by $(2\tau+\frac{1}{2}m^{-1}+4\tau^2m^{-1})$ and then added to (11.5) and the inequality $\lambda^{-2} \geq 1/4\tau^2$ is used on the resulting term in $|\phi|^2$ to yield

$$\int\lambda^{-2m}\{|\phi|^2+\|\phi\|^2\}dt \leq$$

$$\leq (2\tau^2+2\tau+\tfrac{1}{2}m^{-1}+4\tau^2m^{-1})\int\lambda^{-2m}|L\phi-\text{grad}\tilde{p}|^2dt.$$

The lemma follows easily since $\tau < 1$.

The next theorem establishes solutions to P are unique backward in time.

<u>Theorem 11.1</u> (Serrin [169]). Let v and v* be classical solutions to P on $\Omega\times(-T,0)$ with $v = v^*$ on $\Gamma\times(-T,0]$. If $v(x,0) \equiv v^*(x,0)$, then $v \equiv v^*$ in $\Omega\times(-T,0]$.

Proof. Define $u = v^*-v$. Then, if $\tilde{p} = p^*-p$, the pressures corresponding to v* and v, from (10.1) u satisfies

$$u_{i,t} + v_j^*u_{i,j} + u_jv_{i,j} + \tilde{p}_{,i} - \Delta u_i = 0. \qquad (11.7)$$

Therefore,

$$|Lu-grad\tilde{p}|^2 = \int_\Omega (v_j^* u_{i,j} + u_j v_{i,j})(v_r^* u_{i,r} + u_r v_{i,r})dx$$
$$\leq 2M^2(|u|^2 + \|u\|^2), \tag{11.8}$$

where

$$M^2 = \max\{\sup_{\bar{\Omega}\times[-\tau,0]} v_i^* v_i^*, \sup_{\bar{\Omega}\times[-\tau,0]} v_{i,j} v_{i,j}\}.$$

Note that due to the assumed regularity of v, v*, M^2 is finite.

Introduce any C^∞ function $\theta(t)$ which has the behaviour shown in the following diagram:

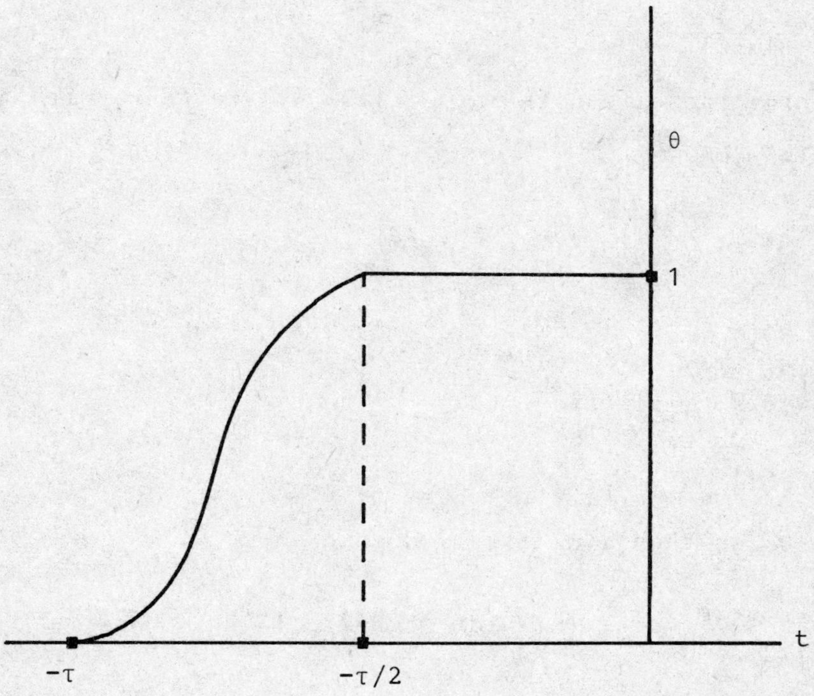

Figure 11.1

The function $\phi = \theta u$ satisfies the conditions of the lemma and so from inequality (11.3) we find

$$\int \lambda^{-2m}(|\phi|^2 + \|\phi\|^2)dt \leqq c\int_{-\tau}^{-\tau/2} \lambda^{-2m}|L\phi - \text{grad}\tilde{p}|^2 dt$$

$$+ c\int_{-\tau/2}^{0} \lambda^{-2m}|Lu - \text{grad}\tilde{p}|^2 dt. \qquad (11.9)$$

Since v, v* are classical solutions, on $(-\tau,-\tau/2)$ we know that

$$|L\phi - \text{grad}\tilde{p}|^2 \leqq A \text{ (constant)}.$$

Furthermore,

$$\int_{-\tau}^{-\tau/2} \lambda^{-2m} dt \leqq \frac{3\tau}{(4m-2)}[3\tau/2]^{-2m}.$$

Define

$$D = \frac{3\tau A}{(4m-2)}.$$

Then, putting these results into (11.9) we see that, with the aid of (11.8),

$$\int \lambda^{-2m}(|\phi|^2 + \|\phi\|^2)dt \leqq cD(2/3\tau)^{2m}$$

$$+ 2cM^2\int_{-\tau/2}^{0} \lambda^{-2m}(|u|^2 + \|u\|^2)dt. \qquad (11.10)$$

Choose now τ so small and m so large that

$$4cM^2 < 1.$$

From (11.10) we then immediately deduce

$$\tfrac{1}{2}\int_{-\tau/2}^{0} \lambda^{-2m}(|u|^2 + \|u\|^2)dt \leqq cD[2/3\tau]^{2m}.$$

Now, restrict the left hand side integral to $(-k\tau,0)$, for any k with $0 < k < \frac{1}{2}$ (Serrin actually chooses $k = 1/4$). Since $\lambda = t-\tau$, $\lambda^{-2m} \geqq [(k+1)\tau]^{-2m}$, and so we may conclude

$$\int_{-k\tau}^{0} (|u|^2 + \|u\|^2)dt \leq 2cD[2(k+1)/3]^{2m}.$$

If the theorem were not true then $|u|^2 > 0$ for some set of values of $t \in (-T,0)$ and we can without loss of generality take this set to be contained in $(-k\tau,0)$. Therefore, since $2(k+1)/3 < 1$, by allowing $m \to \infty$ a contradiction is obtained and the theorem follows.

Remarks.

1. As Serrin [169] points out, the above theorem, together with the corresponding result forward in time shows that a non-constant classical solution to the boundary-initial value problem P can never come to rest in a finite time. This fact is important as this is not always the case for nonlinear partial differential equations, as we saw in chapter 7.
2. An alternative proof of backward uniqueness for P was given by Knops & Payne [90] and for the same problem for the conduction-diffusion Boussinesq equations by the writer [172]. Both of these papers used the technique of logarithmic convexity and were additionally concerned with the question of continuous dependence on the final data.

12 Galdi and Rionero's version of the weighted energy method: an unbounded spatial region

This § studies the boundary-initial value problem P_E on $\Omega\times(0,T)$, $T < \infty$. The object is to obtain a uniqueness theorem without prescribing decay restrictions on the solution and its gradient as $r = (x_i x_i)^{\frac{1}{2}} \to \infty$. In addition, with solutions to (10.1), it is necessary to restrict the behaviour of p, by imposing a suitable decay restriction as $r \to \infty$, and ideally we should like this as weak as possible.

As we shall be concerned with the uniqueness question, it is sufficient to consider equation (11.7) for the difference solution $(\tilde{u},\tilde{p})$. The first proof of uniqueness to P_E is due to D. Graffi [60] who required the solution and its gradient to be bounded and $\tilde{p} = O(1/r)$ at infinity. The pressure condition was weakened to $O(1/r^{\frac{1}{2}+\varepsilon})$, for any $\varepsilon(>0)$, by Serrin (see Edmunds [26]) and further weakened by Cannon & Knightly [12] to $O(1/r^{\frac{1}{2}})$. All of these writers assume the solution and its gradient remain bounded and employ a variant of Graffi's method. The present writer, applying Graffi's method to the conduction-diffusion Boussinesq equations, showed that uniqueness continued to hold when the solution grows like r at infinity, provided similar pressure restrictions hold.

A completely different approach was taken by Galdi & Rionero, who in a series of papers [41,42,163,164] weakened the pressure condition in various ways, one such being $\tilde{p} = O([\log r]^{\sigma}/r^{\frac{1}{2}})$ for any σ, $0 < \sigma < \frac{1}{2}$, and established uniqueness even when the solution behaved like r at infinity and its gradient like r^k for any k (>0). A further approach is due to Fabrizio [28] who required boundedness of the solution but only needed u_i or $\tilde{p}$ to belong to $L^s(\Omega)$ for some $s > 1$. Fabrizio's proof is very interesting as he reduces the uniqueness question to one of establishing existence of a solution to a linear equation.

We are interested in a weighted energy approach, however, and shall describe the method of Galdi & Rionero. For the first demonstration we choose a simple weight function and do not attempt to obtain the best possible result in order that technical details will not obscure the basic idea.

Let as in §7 Ω_o be a bounded, spatial domain of R^3, with boundary Σ and take $\Omega = R^3 - \bar{\Omega}_o$. Let again $B(x,\varepsilon)$ denote the open ball centre x, radius ε, and let $S(x,\varepsilon)$ denote the closed surface surrounding $B(x,\varepsilon)$. Suppose that R_o is the first element of R such that $B(0,R_o) \supset \Omega_o$, and for convenience assume $R_o \geq 1$. Then, for $R \geq R_o$, define

$$\Omega_R = B(0,R) - \bar{\Omega}_o, \tag{12.1}$$

and

$$\Gamma_R = S(0,R), \tag{12.2}$$

cf. figure 7.1. We shall also use the notation

$$\begin{aligned} |u(t)|_1^2 &= \int_\Omega g u_i(\underset{\sim}{x},t) u_i(\underset{\sim}{x},t)\,dx, \\ \|u(t)\|^2 &= \int_\Omega g u_{i,j}(\underset{\sim}{x},t) u_{i,j}(\underset{\sim}{x},t)\,dx, \end{aligned} \tag{12.3}$$

for functions u with values in R^3, where in this § the weight function g is

$$g = e^{-\alpha r}, \quad \alpha > 0. \tag{12.4}$$

The following conditions are imposed on v, v* and $\tilde{p}$:

i. $|v|, |v^*| \leq D$, for each (x,t) in $\Omega\times[0,T]$ and for D ($<\infty$) a positive constant.

ii. $|v_{i,j}|, |v^*_{i,j}|$ may be unbounded at infinity, but grow at most like r^k, for any k (>0). (12.5)

iii. the pressure difference satisfies Serrin's condition, $|\tilde{p}| \leq B/r^{\frac{1}{2}+\varepsilon}$, for some positive constants B, ε, with $0 < \varepsilon < 1$.

<u>Theorem 12.1</u> (cf. Galdi & Rionero [41,163]). Let (v,p), (v*,p*) be two classical solutions to (10.1) in $\Omega\times(0,T)$ and suppose $u = v^*-v$ is zero on $\Sigma\times[0,T)$ and on $\Omega\times\{0\}$. If conditions (12.5) are satisfied then $v \equiv v^*$ in $\Omega\times[0,T)$.

Proof. Commence by multiplying (11.7) by gu_i and integrate over Ω_R to find

$$\frac{d}{dt}\left(\tfrac{1}{2}\int_{\Omega_R} gu^2dx\right)+\int_{\Omega_R} gv^*_ju_iu_{i,j}dx+\int_{\Omega_R} gu_iu_jv_{i,j}dx$$
$$-\int_{\Omega_R} gu_i\Delta u_idx+\int_{\Omega_R} g\tilde{p}_{,i}u_idx = 0. \tag{12.6}$$

Next, denote the second to fifth terms in this equation by I_1-I_4 and after some integrations by parts observe that

$$I_1 = \tfrac{1}{2}\int_{\Sigma\cup\Gamma_R} gn_iv^*_iu^2dS - \tfrac{1}{2}\int_{\Omega_R} g_{,j}v^*_ju^2dx,$$

$$I_2 = \int_{\Sigma\cup\Gamma_R} gn_ju_ju_iv_idS-\int_{\Omega_R} g_{,j}u_jv_iu_idx-\int_{\Omega_R} gu_jv_iu_{i,j}dx,$$

$$I_3 = \tfrac{1}{2}\int_{\Sigma\cup\Gamma_R} n_i g_{,i} u^2 dS - \int_{\Sigma\cup\Gamma_R} g n_j u_i u_{i,j} dS$$
$$+\int_{\Omega_R} g u_{i,j} u_{i,j} dx - \tfrac{1}{2}\int_{\Omega_R} (\Delta g) u^2 dx,$$
$$I_4 = \int_{\Sigma\cup\Gamma_R} n_i u_i \tilde{p} g dS - \int_{\Omega_R} g_{,i} u_i \tilde{p} dx.$$

Since $u \equiv 0$ on Σ, all the integrals over Σ vanish.

The weight function $g = e^{-\alpha r}$ is such that as $R \to \infty$ the integrals over Γ_R will vanish since (12.5) hold. However, unlike §1 where with the heat equation α was chosen in such a way as to allow the solution to grow exponentially in r, the presence of the pressure term necessitates choosing α arbitrarily small and thus we can only allow polynomial growth in $u_{i,j}$. The fact that it is necessary to bound v_i and v_i^* is due to the non-linear term $v_j v_{i,j}$ in the Navier-Stokes equations. Another good feature of the present choice of weight function is that $g_{,i}$ and Δg have simple forms and it is easily verified that

$$\begin{aligned} |g_{,i}| &\leqq \alpha g, \\ |\Delta g| &\leqq \alpha^2 g. \end{aligned} \qquad (12.7)$$

Expressions (12.7) are used in I_1-I_4 which are then put into (12.6) and after allowing $R \to \infty$ we may arrive at

$$\frac{dE}{dt} \leqq \tfrac{1}{2}\alpha\int_\Omega g|v^*|u^2 dx + \alpha\int_\Omega g|v|u^2 dx + \int_\Omega g v_i u_j u_{i,j} dx$$
$$- \|u\|^2 + \tfrac{1}{2}\alpha^2 |u|_1^2 + \alpha\int_\Omega g\, |u|\, |\tilde{p}|\, dx,$$

where

$$E(t) = \tfrac{1}{2}\int_{\Omega} gu^2 dx. \tag{12.8}$$

The bound (12.5,i) is employed on the first three terms on the right of this expression and inequality (AG) is used on the third and sixth terms to obtain

$$\frac{dE}{dt} \leqq ME + \tfrac{1}{2}\alpha^2 |\tilde{p}|_1^2 \ , \tag{12.9}$$

where

$$M = 3\alpha + \tfrac{1}{2}D^2 + \alpha^2 + 1.$$

To deal with the pressure term use (12.5,iii) as follows;

$$\tfrac{1}{2}\alpha^2 |\tilde{p}|_1^2 \leqq \tfrac{1}{2}B\alpha^2 \int_0^{2\pi} d\theta \int_0^{\pi} \sin\phi d\phi \int_1^{\infty} \frac{r^2 e^{-\alpha r} dr}{r^{1+2\varepsilon}}$$

$$= 2\pi B\alpha^{2\varepsilon} \int_{\alpha}^{\infty} e^{-s} s^{1-2\varepsilon} ds.$$

Since $\varepsilon < 1$ the integral is bounded for each $\alpha \geqq 0$ and so if we denote by k the coefficient of $\alpha^{2\varepsilon}$ then from (12.9) we see that

$$\frac{dE}{dt} \leqq ME + k\alpha^{2\varepsilon}. \tag{12.10}$$

Integrate this from 0 to t and note that $E(0)=0$ so that

$$E(t) \leqq e^{Mt} k\alpha^{2\varepsilon}/M. \tag{12.11}$$

Next, observe that for any $R \geqq R_o$,

$$E(t) \geqq \tfrac{1}{2}e^{-\alpha R} \int_{\Omega_R} u^2 dx, \tag{12.12}$$

and so combining (12.11) and (12.12) together, for $0 \leqq t \leqq T$, we find

$$\int_{\Omega_R} u^2 dx \leq \frac{2k}{M}.\exp(\alpha R+MT).\alpha^{2\varepsilon}. \qquad (12.13)$$

If $u \neq 0$ anywhere in $\Omega\times\{t\}$ we choose R (fixed) so that the left hand side is positive. Then, by allowing $\alpha \to 0$ a contradiction is obtained. The number R is arbitrary and so $u \equiv 0$ in $\Omega\times[0,T)$ and the proof of the theorem is complete.

13 Backward uniqueness for solutions to the Navier-Stokes equations on an exterior domain

The backward uniqueness result of Serrin described in §11 was extended to the situation in which the spatial domain is unbounded, without requiring stringent decay hypotheses at infinity by Edmunds [26] who employed a variant of the method of D. Graffi In this § we study the question addressed by Edmunds [26], although the approach adopted is a weighted energy one in both space and time. In this way, by combining the ideas of Serrin [169], Edmunds [26] and Rionero & Galdi [163] we are able to show that a solution to P_E on $\Omega\times(-T,0)$ is unique, under weaker decay assumptions than those required in Edmunds' analysis.

The notation to be employed here is that of §12, except a different weight function, g, is selected and different decay or growth conditions are imposed to those of (12.5). To state the necessary conditions let v_i, v_i^* be classical solutions to P_E on $\Omega\times(-\infty,0)$ which satisfy the same boundary conditions on $\Sigma\times(-\infty,0)$ and are such that $\underset{\sim}{v} \equiv \underset{\sim}{v}^*$ when $t = 0$. Define $u_i = v_i^* - v_i$ and suppose:

$|v^*|$ and $|v_{i,j}|$ are uniformly bounded by a constant $D\ (<\infty)$, for each x,t in $\Omega\times(-\infty,0]$. Additionally, $|v|$ may grow with respect to r like r and $|v^*_{i,j}|$, $|u_{i,t}|$ like r^k for any k (>0). (13.1)

The difference pressure $\tilde{p}$, corresponding to u_i, satisfies the condition

$$|\tilde{p}| \leq M^{\frac{1}{2}}(\log r)^{\sigma}/r^{\frac{1}{2}},$$

for some constant M (>0) and some $\sigma\in[0,\frac{1}{2})$. (13.2)

$$\int_{\Omega} u_{i,j}u_{i,j}\,dx \quad \text{exists for each } t < 0. \tag{13.3}$$

The pressure condition (13.2) was introduced by Rionero & Galdi [163], and we employ it here instead of the stronger condition of Serrin (12.5,iii) which was used by Edmunds.

Theorem 13.1. Let v_i and v_i^* be classical solutions to P_E on $\Omega\times(-\infty,0)$, which satisfy (13.1)-(13.3). If $u_i(\underset{\sim}{x},0) \equiv 0$ $\forall x\in\Omega$ and $u_i \equiv 0$ $\forall(x,t)\in\Sigma\times(-\infty,0)$, then $v_i \equiv v_i^*$ on $\Omega\times(-\infty,0]$.

Proof. Let τ, m be positive numbers, with m an integer, and define $\lambda = t-\tau$, $\psi_i = \lambda^{-m}u_i$. From (11.7) we find ψ_i satisfies the equation

$$\Delta\psi_i - \psi_{i,t} - m\lambda^{-1}\psi_i - \lambda^{-m}\tilde{p}_{,i} = \lambda^{-m}(v_j^*u_{i,j}-u_jv_{i,j}). \tag{13.4}$$

For positive a, γ, with $a \geq 5$, we choose the weight function g as

$$g(r) = \exp(-ar^{\gamma}). \tag{13.5}$$

Clearly, the exponential has the same effect as the function chosen in §12 regarding dominating the boundary terms for large R. However, the choice (13.5) allows treatment of the weaker pressure condition (13.2).

Multiply (13.4) by $-g\psi_{i,t}$ and integrate over Ω_R for some $R > R_o$, to find after integrating by parts and noting $\psi_i \equiv 0$ on Σ,

$$\frac{1}{2}\frac{d}{dt}\int_{\Omega_R} g\psi_{i,j}\psi_{i,j}dx+\int_{\Omega_R} g\psi_{i,t}\psi_{i,t}dx+\frac{1}{2}m\frac{d}{dt}\int_{\Omega_R} g\lambda^{-1}\psi_i\psi_i dx$$

$$+\frac{1}{2}m\lambda^{-2}\int_{\Omega_R} g\psi_i\psi_i dx = -\int_{\Omega_R} g_{,b}\psi_{i,t}\psi_{i,b}dx+\int_{\Omega_R}\tilde{p}g_{,i}\lambda^{-m}\psi_{i,t}dx$$

$$-\int_{\Omega_R}\lambda^{-m}g\psi_{i,t}(v_j^*u_{i,j}+u_jv_{i,j})dx + \int_{\Gamma_R} gn_b\psi_{i,t}\psi_{i,b}dS$$

$$-\int_{\Gamma_R} g\psi_{i,t}n_i\tilde{p}\lambda^{-m}dS. \qquad (13.6)$$

Next, allow $R \to \infty$ in (13.6). Since conditions (13.1) and (13.2) hold the boundary terms disappear and we arrive at

$$\frac{1}{2}\frac{d}{dt}\,||\psi||^2 +|\psi_t|_1^2+\frac{1}{2}m\frac{d}{dt}(\lambda^{-1}|\psi|_1^2)+\frac{1}{2}m\lambda^{-2}|\psi|_1^2 = -\int_\Omega g_{,b}\psi_{i,t}\psi_{i,b}dx$$

$$+ \int_\Omega \lambda^{-m}\tilde{p}g_{,i}\psi_{i,t}dx - \int_\Omega g\psi_{i,t}(v_j^*\psi_{i,j}+\psi_jv_{i,j})dx \qquad (13.7)$$

This expression is now integrated over $(-\tau,0)$ and since $\psi_i \equiv 0$ for $t = 0$ we obtain

$$-\frac{1}{2}\,||\psi(-\tau)||^2 +\int|\psi_t|_1^2dt+\frac{1}{2}m\int\lambda^{-2}|\psi|_1^2dt+\frac{m}{4\tau}|\psi(-\tau)|_1^2 =$$

$$= -\iint_\Omega g_{,b}\psi_{i,t}\psi_{i,b}dxdt + \iint_\Omega \lambda^{-m}\psi_{i,t}\tilde{p}g_{,i}dxdt$$

$$-\iint_\Omega g\psi_{i,t}(v_j^*\psi_{i,j}+v_{i,j}\psi_j)dxdt, \qquad (13.8)$$

where here and for the remainder of this § an integral over an unspecified domain is understood to be over $(-\tau,0)$.

The next step involves multiplying (13.4) by $-g\psi_i$ and then in a derivation similar to that for (13.8) we may obtain

$$\int ||\psi||^2\,dt \leqq m\int|\lambda^{-1}|\,|\psi|_1^2dt + \frac{1}{2}|\psi(-\tau)|_1^2$$

$$+ \iint_\Omega|g_{,j}\psi_i\psi_{i,j}|dxdt + \frac{1}{2}\iint_\Omega|v_i^*|\,|g_{,j}|\psi_i\psi_i dxdt$$

$$+\iint_{\Omega} g|v_{i,k}\psi_i\psi_k|\,dxdt+\iint_{\Omega}|\tilde{p}||\lambda^{-m}||g_{,i}\psi_i|\,dxdt. \qquad (13.9)$$

We now multiply (13.8) by $8\tau^2m^{-1}+m^{-1}+4\tau$ and add this to (13.9) and from the result deduce that,

$$(4\tau+m^{-1}+8\tau^2m^{-1})\int|\psi_t|_1^2dt+\int\|\psi\|^2\,dt+\tfrac{1}{2}\int\lambda^{-2}|\psi|_1^2dt$$

$$+2\tau m\int\lambda^{-2}|\psi|_1^2dt+4\tau^2\int\lambda^{-2}|\psi|_1^2dt \leq m\int|\lambda^{-1}||\psi|_1^2dt$$

$$+\iint_{\Omega}|g_{,a}\psi_i\psi_{i,a}|\,dxdt+\tfrac{1}{2}\iint_{\Omega}|g_{,a}v_a^*\psi_i\psi_i|\,dxdt+\iint_{\Omega}|\tilde{p}g_{,i}\psi_i\lambda^{-m}|\,dxdt$$

$$+(4\tau^2m^{-1}+\tfrac{1}{2}m^{-1}+2\tau)\,\|\psi(-\tau)\|^2+(8\tau^2m^{-1}+m^{-1}+4\tau)\{\iint_{\Omega}|\psi_{i,t}g_{,a}\psi_{i,a}|\,dxdt$$

$$+\iint_{\Omega}|\lambda^{-m}\psi_{i,t}\tilde{p}g_{,i}|\,dxdt+\iint_{\Omega}g|\psi_{i,t}(v_j^*\psi_{i,j}+v_{i,j}\psi_j)|\,dxdt\}$$

$$+\iint_{\Omega}g|v_{i,j}|\psi_a\psi_a\,dxdt. \qquad (13.10)$$

From (11.6) $|\lambda^{-1}| \leq 2\tau\lambda^{-2}$ and with the aid of this the fourth term on the left of (13.10) may be discarded together with the first term on the right. Moreover, from (11.6) $\lambda^{-2} \geq 1/4\tau^2$ and so the fifth term on the left may be replaced by $\int|\psi|_1^2dt$. The last term in (13.10) is dealt with using inequality (AG), $(11.6)_2$ and the bound (13.1) on $|v_{i,j}|$, as follows;

$$\iint_{\Omega}g|v_{i,j}|\lambda^{-2m}u_au_a\,dxdt \leq \tfrac{1}{2}\int\lambda^{-2m-2}|u|_1^2dt+\frac{D^2}{2}\int\lambda^{-2m+2}|u|_1^2dt,$$

$$\leq \tfrac{1}{2}\int\lambda^{-2}|\psi|_1^2dt+2\tau^2D^2\int|\psi|_1^2dt. \qquad (13.11)$$

Observe that $g_{,i} = -\gamma ar^{\gamma-1}g\dfrac{x_i}{r}$ and so

$$|g_{,i}| \leq \gamma ar^{\gamma-1}g, \qquad (13.12)$$

and if we restrict γ so that $0 < \gamma < 1$, since $r \geq 1$,

$$|g_{,i}| \leqq \gamma a g. \tag{13.13}$$

We also select τ so small that

$$2\tau^2 D^2 \leqq \tfrac{1}{2}, \tag{13.14}$$

and define C by

$$C = 10(\tau + m^{-1}). \tag{13.15}$$

We assume now (13.10) is modified as suggested in the paragraph preceding (13.11) and using (13.11)-(13.15), with τ chosen small enough and m large enough, we derive from (13.10) the following inequality,

$$\begin{aligned}
&\frac{C}{10}\int |\psi_t|_1^2 dt + \tfrac{1}{2}\int |\psi|_1^2 dt + \int \|\psi\|^2\, dt \leqq \frac{C}{2}\|\psi(-\tau)\|^2 \\
&+\iint_\Omega \gamma a g(|\psi_i||\psi_{i,j}| + \tfrac{1}{2}|v_i^*|\psi_b\psi_b)\,dxdt \\
&+a\gamma\iint_\Omega g r^{\gamma-1}|\tilde{p}||\psi_i||\lambda^{-m}|\,dxdt + C\{a\gamma\iint_\Omega g(|\psi_{i,t}||\psi_{i,j}| + \\
&+ r^{\gamma-1}|\tilde{p}\lambda^{-m}||\psi_{i,t}|)dxdt + \iint_\Omega gD(|\psi_{i,t}||\psi_{i,j}| + |\psi_{i,t}||\psi_j|)dxdt\}.
\end{aligned} \tag{13.16}$$

It is now necessary to remove the first term on the right of (13.16) and this is done with the aid of an argument adapted from Edmunds [26].

We shall temporarily denote the dependence on γ in the weighted Dirichlet integral of u by $\|u\|_\gamma$. Then, by (13.3) $\|u\|_0^2$ is finite for $t < 0$ and so we choose γ_1 so that

$$\|u(-\tau)\|_\gamma^2 \geqq \frac{1}{\sqrt{2}}\|u(-\tau)\|_0^2, \quad \forall \gamma \leqq \gamma_1.$$

By continuity, there exists $\delta(\gamma_1)$, $0 < \delta < \tau$, with

$$\|u(t)\|^2_{\gamma_1} \geq \frac{1}{\sqrt{2}} \|u(-\tau)\|^2_{\gamma_1} ,$$

for every t such that $-\tau \leqq t \leqq -\tau+\delta$. Restrict $\gamma \leqq \gamma_1$ and then for $-\tau \leqq t \leqq -\tau+\delta$,

$$\|u(t)\|^2_{\gamma} \geq \|u(t)\|^2_{\gamma_1} \geq \frac{1}{\sqrt{2}} \|u(-\tau)\|^2_{\gamma_1} \geq \tfrac{1}{2} \|u(-\tau)\|^2_0$$
$$\geq \tfrac{1}{2} \|u(-\tau)\|^2_{\gamma} .$$

It follows that

$$\|u(t)\|^2 \geq \tfrac{1}{2} \|u(-\tau)\|^2 , \quad -\tau \leqq t \leqq -\tau+\delta,$$

so

$$\int_{-\tau}^{0} \|\psi\|^2 \, dt \geq \int_{-\tau}^{-\tau+\delta} \lambda^{-2m} \|u\|^2 \, dt \geq \tfrac{1}{2} \|u(-\tau)\|^2 \int_{-\tau}^{-\tau+\delta} \lambda^{-2m} dt.$$

Therefore,

$$\tfrac{1}{2}\int \|\psi\|^2 \, dt - \tfrac{1}{2}C \|\psi(-\tau)\|^2 \geq \tfrac{1}{2} \|u(-\tau)\|^2 (2\tau)^{-2m} \times$$
$$\times \left[\frac{\tau}{(2m-1)}\left((1-\delta/2\tau)^{-2m+1}-1\right)-10(\tau+m^{-1})\right].$$

Provided m is large enough this expression is positive and so we may remove the first term on the right of (13.16) and replace the third term on the left by $\tfrac{1}{2}\int \|\psi\|^2 \, dt$.

The remaining terms on the right of (13.16) are estimated using inequality (AG) in such a way as to arrive at the following inequality,

$$\frac{C}{10}\int |\psi_t|^2_1 dt + \tfrac{1}{2}\int |\psi|^2_1 dt + \tfrac{1}{2}\int \|\psi\|^2 \, dt \leq C\left(\frac{\gamma a}{2} + \frac{a\gamma^{\varepsilon}}{2} + \frac{1}{25}\right)\int |\psi_t|^2_1 dt$$
$$+ \left(\frac{25}{2}CD^2 + [1+C]\frac{\gamma a}{2}\right)\int \|\psi\|^2 \, dt + \left(\frac{25}{2}CD^2 + \frac{a}{2}[\gamma + D\gamma + \gamma^{\varepsilon}]\right)\int |\psi|^2_1 dt$$

$$+ \tfrac{1}{2}a(1+C)\gamma^{2-\varepsilon}\iint_{\Omega}\lambda^{-2m}gr^{2\gamma-2}\tilde{p}^2 dxdt, \tag{13.17}$$

where ε is chosen so that $0 < \varepsilon < 1-2\sigma$. The numbers τ, m, γ are now selected so that

$$\frac{1}{10} \geqq \tfrac{1}{2}a(\gamma+\gamma^{\varepsilon}) + \frac{1}{25},$$

$$\frac{1}{2} \geqq \frac{25CD^2}{2} + \frac{\gamma a}{2}(1+C),$$

$$\frac{1}{4} \geqq \frac{25CD^2}{2} + \tfrac{1}{2}a(\gamma+D\gamma+\gamma^{\varepsilon}),$$

and then from (13.17) we find

$$\int|\psi|_1^2 dt \leqq 2a(1+C)\gamma^{2-\varepsilon}\iint_{\Omega} g\lambda^{-2m}r^{2\gamma-2}\tilde{p}^2 dxdt. \tag{13.18}$$

The pressure difference $\tilde{p}$ satisfies (13.2) and use of this in (13.18) leads to

$$\int_{-\tau}^{0}\int_{\Omega} g\lambda^{-2m}u_i u_i dxdt \leqq 8\pi Ma(1+C)\gamma^{2-\varepsilon}\left[\int_{-\tau}^{0}\lambda^{-2m}dt\right]\times$$
$$\times\left[\int_1^{\infty} r^{2\gamma-1}\exp(-ar^{\gamma})(\log r)^{2\sigma}dr\right]. \tag{13.19}$$

By direct calculation

$$\int_{-\tau}^{0}\lambda^{-2m}dt \leqq \frac{\tau}{(2m-1)}\tau^{-2m}. \tag{13.20}$$

Moreover, for $a \geqq 5$,

$$I_1 = \int_1^{\infty} r^{2\gamma-1}e^{-ar^{\gamma}}(\log r)^{2\sigma}dr \leqq \int_1^{\infty}\frac{e^{-2r^{\gamma}}(\log r)^{2\sigma}}{r^{\gamma+1}}dr. \tag{13.21}$$

The maximum of $\exp(-r^{\gamma})(\log r)^{2\sigma}$ occurs where $\gamma r_m^{\gamma} = 2\sigma/\log r_m$, and so since $r \geqq 1$, $\log r_m \leqq 2\sigma/\gamma$. Hence,

$$I_1 \leqq \frac{(2\sigma)^{2\sigma}}{\gamma^{2\sigma}} \int_1^\infty \frac{e^{-r^\gamma} dr}{r^{\gamma+1}} \leqq \frac{(2\sigma)^{2\sigma}}{e} \gamma^{-1-2\sigma}. \tag{13.22}$$

The bounds (13.20) and (13.22) are now inserted into (13.19) to derive

$$\int_{-\tau}^0 \int_\Omega g\lambda^{-2m} u_i u_i \, dx dt \leqq k\gamma^{1-2\sigma-\varepsilon} \tau^{-2m}, \tag{13.23}$$

where

$$k = \frac{8\pi Ma(1+C)\tau(2\sigma)^{2\sigma}}{(2m-1)e}.$$

To establish the theorem, suppose the conclusion false. Then, without loss of generality we may suppose that for some finite R, $u_i \neq 0$ on $\Omega_R \times (-\tau, 0)$. Define $\beta = 1-2\sigma-\varepsilon$ and note that $\beta > 0$. From (13.23) we now easily deduce

$$\int_{-\tau}^0 \int_{\Omega_R} u_i u_i \, dx dt \leqq k\exp(aR^\gamma)\gamma^\beta 2^{2m}. \tag{13.24}$$

Finally, consistent with our earlier use of γ and m, choose $\gamma = (1/3)^{2m/\beta}$ and from (13.24) we see that

$$0 < \int_{-\tau}^0 \int_{\Omega_R} u_i u_i \, dx dt \leqq L(2/3)^{2m}, \tag{13.25}$$

where $L = k\exp(aR^\gamma)$. By allowing $m \to \infty$ a contradiction is obtained for any (fixed) value of R and the theorem follows.

14 Unique continuation for solutions to the Navier-Stokes equations on an exterior domain

The nature of the work presented in this § is to try to determine the rate of decay in time of solutions to the Navier-Stokes equations. An indirect approach is adopted, whereby it is established that if solutions were to have a certain type of behaviour as $t \to \infty$, such solutions would necessarily be zero. This approach, for parabolic differential equations and inequalities, is due to Protter [154]; the ideas appropriate to the Navier-Stokes equations were derived by Edmunds [27].

As Protter [154] points out, the starting point of the investigation of asymptotic behaviour is the knowledge that solutions to the heat equation on a bounded domain, with homogeneous boundary data, decay like $e^{-\lambda t}$ for some positive λ as $t \to \infty$. It is natural to wonder if all solutions to the Navier-Stokes equations enjoy such properties. For a bounded domain, Ω, Edmunds [27] employed a weighted energy argument similar to that of Protter [154] to show that if a classical solution to the Navier-Stokes equations (10.1), with a conservative body force, vanishes on the boundary of the spatial region and is such that as $t \to \infty$, $\sup_{\bar{\Omega}} |v_{i,j}| = O(e^{-\mu t})$ for every $\mu (>0)$, then $v_i \equiv 0$ on $\Omega \times [0,\infty)$. Note that this result does not impose any condition on $v_i(x,0)$ and is thus a unique continuation result in the sense that any solution satisfying the required decay estimate must be continued as the zero solution backward in time.

Edmunds [27] also showed how a similar result may be established when Ω is an exterior region by combining Protter's weighted energy arguments with the Graffi technique. The main contribution of this § is to derive a similar result to that of Edmunds but by using only weighted energy arguments; essentially by combining the methods of Protter, Edmunds and Rionero & Galdi.

We shall employ the notation of §§12 and 13 and so let (v_i, p) be a classical solution to P_E on $\Omega\times I$. Here $I = [0,\infty)$ and we suppose the body force is conservative and so may be incorporated in the pressure term. A solution is said to be of class K, if:

$$V(t) = \sup_{x\in\Omega^-} \max_{i,j=1,2,3} \{|v_i|, |v_{i,t}|, |v_{i,j}|\} \tag{14.1}$$

is a bounded function and $V(t) = o(e^{-\mu t})$, as $t \to \infty$, for each μ (>0).

$$||v(t)||_0^2 = \int_\Omega v_{i,j}(x,t)v_{i,j}(x,t)dx \quad \text{is finite for every}$$

$t \in I$. (14.2)

$$|p| \leq A(t)\frac{(\log r)^\sigma}{r^{\frac{1}{2}}}, \tag{14.3}$$

for some $\sigma\in(0,\frac{1}{2})$. $A(t)$ is bounded on I and $A = o(e^{-\mu t})$, as $t \to \infty$, for each μ (>0).

Either,

(i) $$|v_{i,j}| \leq A(t)\frac{(\log r)^\sigma}{r^{\frac{1}{2}}}, \tag{14.4}$$

where σ, $A(t)$ are as in (14.3), or

(ii) $||v(t)|| \leq A(t)$.

$$|v(t)|_1,\ |v_t(t)|_1 \leq A(t)p(\gamma), \tag{14.5}$$

where $p(\gamma)$ is a smooth function for $\gamma > 0$, but possibly $p \to \infty$ as $\gamma \to 0$.

Conditions (14.1) and (14.2) were imposed by Edmunds, and instead of (14.4) he required $\int_{\Gamma_R} v_{i,j}v_{i,j}dS = A(t)r^{\frac{1}{2}-\varepsilon}$, for some ε (>0). However, while (14.3) is weaker than the corresponding condition of Edmunds, we find it necessary to impose in addition (14.5). Since the germ of the Galdi and Rionero approach is to allow growth behaviour in r by weighting the integrand with a decaying exponential, (14.2) is a somewhat strong condition, and with respect to this and (14.5) I feel it should be possible to establish a stronger result than theorem 14.1, by use of the same techniques.

Theorem 14.1. Let (v_i,p) be a classical solution to P_E of class K and suppose $v_i(x,t) \equiv 0$ on $\Sigma\times I$. Then $v_i \equiv 0$ in $\Omega\times I$.

Proof. Define $z_i = t^{-\beta}e^{\lambda t}v_i$, for positive constants β, λ to be suitably chosen. In terms of z_i, (10.1) for $f_i = -\phi_{,i}$, some ϕ, may be written with $p \equiv p+\phi$,

$$\Delta z_i - z_{i,t} + z_i(\lambda-\beta t^{-1}) - t^{-\beta}e^{\lambda t}p_{,i} = t^{-\beta}e^{\lambda t}v_k v_{i,k}. \tag{14.6}$$

Now multiply this expression by $z_{i,t}g$, where g is given by (13.5), integrate over Ω_R and then allow $R \to \infty$, after some integration by parts, to derive

$$-\frac{d}{dt}\tfrac{1}{2}\|z\|^2 - \int_\Omega g_{,k}z_{i,t}z_{i,k}dx - |z_t|_1^2 - \frac{\beta}{2t^2}|z|_1^2 + \frac{d}{dt}\{\tfrac{1}{2}(\lambda-\beta t^{-1})|z|_1^2\}$$

$$= -\int_\Omega g_{,i}z_{i,t}pt^{-\beta}e^{\lambda t}dx + \int_\Omega gz_{i,t}v_j v_{i,j}t^{-\beta}e^{\lambda t}dx.$$

We next select $T_o(>0)$ and integrate this equation from T_o to ∞. Since (v_i,p) was chosen of class K, the result is

$$\int |z_t|_1^2 dt + \tfrac{1}{2}\beta\int \frac{dt}{t^2}|z|_1^2 - \tfrac{1}{2}\,\|z(T_o)\|^2 + \tfrac{1}{2}(\lambda-\beta/T_o)|z(T_o)|_1^2 =$$

$$= -\iint_\Omega g_{,k}z_{i,t}z_{i,k}dxdt + \iint_\Omega t^{-\beta}e^{\lambda t}pg_{,i}z_{i,t}dxdt$$

$$-\iint_\Omega t^{-\beta}e^{\lambda t}gz_{i,t}v_jv_{i,j}dxdt, \qquad (14.7)$$

where all integrals are over (T_o,∞) unless otherwise indicated.

Next, multiply (14.6) by gz_i, integrate over Ω_R, integrate by parts on the Δz_i and $p_{,i}$ terms, and then allow $R \to \infty$ to obtain

$$-\|z\|^2 - \int_\Omega z_iz_{i,k}g_{,k}dx - \tfrac{1}{2}\frac{d}{dt}|z|_1^2 + (\lambda-\beta t^{-1})|z|_1^2$$

$$= -t^{-\beta}e^{\lambda t}\int_\Omega pz_ig_{,i}dx + t^{-\beta}e^{\lambda t}\int_\Omega gz_iv_kv_{i,k}dx.$$

We employ inequality (AG) on the last term and then integrate the result over (T_o,∞) to find

$$\int \|z\|^2\, dt \leqq \tfrac{1}{2}|z(T_o)|_1^2 + \tfrac{1}{2}\int t^{1-2\beta}e^{2\lambda t}\int_\Omega |v_jv_{i,j}|^2 gdxdt$$

$$+\int(\lambda-\beta t^{-1}+\tfrac{1}{2}t^{-1})|z|_1^2dt$$

$$+\iint_\Omega g_{,i}z_ipt^{-\beta}e^{\lambda t}dxdt - \iint_\Omega z_iz_{i,j}g_{,j}dxdt. \qquad (14.8)$$

Take now $\beta = \lambda+1$ in (14.8) and choose λ, $T_o > 1$. With $w_i = z_it$, i.e. $w_i = t^{-\lambda}e^{\lambda t}v_i$, we may then deduce from (14.8)

$$\int\frac{dt}{t^2}\|w\|^2 \leqq \frac{1}{2T_o^2}|w(T_o)|_1^2 + \iint_\Omega \tfrac{1}{2}gv_rv_rw_{i,j}w_{i,j}dxdt$$

$$+\lambda\int\frac{dt}{t^2}|w|_1^2 + \iint_\Omega e^{\lambda t}t^{-\lambda-1}z_ig_{,i}pdxdt - \iint_\Omega g_{,k}z_iz_{i,k}dxdt\ ,$$

$$\leq \frac{1}{2T_o^2}|w(T_o)|_1^2 + \tfrac{1}{2}\iint_\Omega g v_r v_r w_{i,k} w_{i,k}\,dxdt + \tfrac{1}{2}a\gamma\int \frac{dt}{t^2}\,\|w\|^2$$

$$+\left(\frac{5\lambda}{4}+\frac{a\gamma}{2}\right)\int \frac{dt}{t^2}|w|_1^2 + \frac{a^2\gamma^2}{\lambda}\iint_\Omega g p^2 t^{-2(1+\lambda)} e^{2\lambda t} r^{2(\gamma-1)}\,dxdt, \quad (14.9)$$

where in the last line inequalities (AG) and (13.12) have been used.

Now choose $\beta = \lambda$ in (14.7) (so that $w_i = z_i$ there) and then from that inequality with the aid of (AG) and (13.12) we derive

$$\int |w_t|_1^2 dt + \frac{\lambda}{2}\int \frac{dt}{t^2}|w|_1^2 + \tfrac{1}{2}\lambda(1-1/T_o)|w(T_o)|_1^2 \leq$$

$$\leq \tfrac{1}{2}\,\|w(T_o)\|^2 + \tfrac{1}{2}\int |w_t|_1^2 dt + a^2\gamma^2\iint_\Omega g p^2 t^{-2\lambda} e^{2\lambda t} r^{2(\gamma-1)}\,dxdt$$

$$+ 2\iint_\Omega g t^{-2\lambda} e^{2\lambda t} v_j v_{i,j} v_m v_{i,m}\,dxdt + 2a^2\gamma^2\iint_\Omega g r^{2(\gamma-1)} w_{i,j} w_{i,j}\,dxdt. \quad (14.10)$$

Inequality (14.10) is now multiplied by 4 and added to (14.9) to obtain

$$2\int |w_t|_1^2 dt + \left[\frac{3\lambda}{4} - \tfrac{1}{2}a\gamma\right]\int \frac{dt}{t^2}|w|_1^2 + (1-\tfrac{1}{2}a\gamma)\int \frac{dt}{t^2}\,\|w\|^2$$

$$+ \left[2\lambda(1-1/T_o) - 1/2T_o^2\right]|w(T_o)|_1^2 \leq 2\,\|w(T_o)\|^2$$

$$+\frac{17}{2}\iint_\Omega g v_r v_r w_{i,j} w_{i,j}\,dxdt + \gamma^2 a^2\iint_\Omega \left(4+\frac{1}{\lambda t^2}\right) g p^2 t^{-2\lambda} e^{2\lambda t} r^{2(\gamma-1)}\,dxdt$$

$$+ 8a^2\gamma^2\iint_\Omega g r^{2(\gamma-1)} w_{i,j} w_{i,j}\,dxdt. \quad (14.11)$$

We now restrict γ, λ, T_o so that $\lambda > 1$, $T_o > 2$, $a\gamma \leq \frac{1}{2}$ and from (14.11) we are able to deduce

$$2\int |w_t|_1^2 dt + \tfrac{1}{2}\lambda\int \frac{dt}{t^2}|w|_1^2 + \frac{3}{4}\int \frac{dt}{t^2}\,||w||^2 \leq 2\,||w(T_o)||^2$$

$$+ \frac{17}{2}\iint_\Omega g v_r v_r w_{m,n} w_{m,n}\,dxdt + 5a^2\gamma^2\iint_\Omega g p^2 t^{-2\lambda} e^{2\lambda t} r^{2(\gamma-1)}\,dxdt$$

$$+ 8a^2\gamma^2\iint_\Omega g r^{2(\gamma-1)} w_{i,j} w_{i,j}\,dxdt. \qquad (14.12)$$

Since v_i satisfies (14.1) we further select T_o so large that

$$\frac{17}{2}\int_\Omega g v_k v_k w_{i,j} w_{i,j}\,dx \leq \frac{1}{4t^2}\,||w||^2, \quad \text{for all } t \geq T_o. \qquad (14.13)$$

Using (14.13) in (14.12) yields

$$2\int |w_t|_1^2 dt + \tfrac{1}{2}\lambda\int \frac{dt}{t^2}|w|_1^2 + \tfrac{1}{2}\int \frac{dt}{t^2}\,||w||^2 \leq$$

$$\leq 2\,||w(T_o)||^2 + 8a^2\gamma^2\iint_\Omega g t^{-2\lambda} e^{2\lambda t} r^{2(\gamma-1)}[p^2 + v_{i,j} v_{i,j}]\,dxdt. \qquad (14.14)$$

To remove the first term on the right of (14.14) we adapt an argument due to Edmunds [27]. Let us again emphasize the dependence of γ as in the argument following (13.16). By (14.2) $||v||_0^2$ is finite for all t in I and so we select γ_1 such that

$$||v(T_o)||_\gamma^2 \geq \tfrac{1}{2}\,||v(T_o)||_0^2, \quad \text{for all } \gamma \leq \gamma_1.$$

By continuity there exists $\delta(\gamma_1)(>0)$ such that

$$||v(t)||_{\gamma_1}^2 \geq \tfrac{1}{2}\,||v(T_o)||_{\gamma_1}^2,$$

for $T_o \leq t \leq T_o+\delta$. Therefore, for $T_o \leq t \leq T_o+\delta$ and $\gamma \leq \gamma_1$, we have

$$||v(t)||_\gamma^2 \geq ||v(t)||_{\gamma_1}^2 \geq \tfrac{1}{2}||v(T_o)||_{\gamma_1}^2 \geq \frac{1}{4}||v(T_o)||_0^2 \geq \frac{1}{4}||v(T_o)||_\gamma^2.$$

Hence

$$\tfrac{1}{2}\int\frac{dt}{t^2}\,||w||^2 \;-\; 2\,||w(T_o)||^2 \;\geqq$$

$$\geqq \frac{1}{4}\,||v(T_o)||^2 \int_{T_o}^{T_o+\delta} t^{-2(\lambda+1)}e^{2\lambda t}dt - 2\,||v(T_o)||^2\, T_o^{-2\lambda}e^{2\lambda T_o},$$

$$\geqq \frac{||v(T_o)||^2}{4}\{(T_o+\delta)^{-2(\lambda+1)}\int_{T_o}^{T_o+\delta} e^{2\lambda t}dt - 8T_o^{-2\lambda}e^{2\lambda T_o}\},$$

$$= \frac{||v(T_o)||^2\, e^{2\lambda T_o}(T_o+\delta)^{-2(\lambda+1)}}{8\lambda}\{e^{2\lambda\delta}-1-16\lambda(T_o+\delta)^2[1+\delta/T_o]^{2\lambda}\}.$$

For λ large enough this term is positive and hence (14.14) may be reduced to

$$\tfrac{1}{2}\lambda\int dt.e^{2\lambda t}t^{-2(\lambda+1)}\,|v|_1^2 \leq$$

$$\leq 8a^2\gamma^2\int t^{-2\lambda}e^{2\lambda t}\int_\Omega gr^{2(\gamma-1)}[p^2+v_{i,j}v_{i,j}]dxdt. \qquad (14.15)$$

It remains to estimate p^2 and $v_{i,j}v_{i,j}$. Suppose (14.4,i) holds. Then, as p satisfies (14.3) the right hand side of (14.15) may be bounded by

$$16a^2\gamma^2[\int_{T_o}^{\infty} A^2(t)t^{-2\lambda}e^{2\lambda t}dt][\int_\Omega e^{-ar^\gamma}r^{2\gamma-1}(\log r)^{2\sigma}dx]. \qquad (14.16)$$

By virtue of the properties of A(t) the term involving this is a constant. The integral over Ω is handled exactly as in the steps following (13.18). We use the result in (14.15) to obtain

$$\lambda\int e^{2\lambda t}t^{-2(\lambda+1)}\,|v|_1^2dt \leq Q\gamma^{1-2\sigma}, \qquad (14.17)$$

where

$$Q = 32a^2[\int_{T_o}^{\infty} A^2(t)t^{-2\lambda}e^{2\lambda t}dt](2\sigma)^{2\sigma}e^{-2\sigma}. \qquad (14.18)$$

The proof now proceeds by contradiction. For, if the theorem is false then for some R, $v_i \not\equiv 0$ in $\Omega_R\times(T_o,\infty)$. So, from (14.17) we deduce

$$0 < \lambda\iint_{\Omega_R} e^{2\lambda t}t^{-2(\lambda+1)}v_iv_i dxdt \leqq Qe^{aR^\gamma}.\gamma^{1-2\sigma}. \qquad (14.19)$$

A contradiction is immediately obtained for small enough γ. Thus $v_i \equiv 0$ on $\Omega_R\times(T_o,\infty)$. However, R is arbitrary beyond some lower value and so $v_i \equiv 0$ on $\Omega\times(T_o,\infty)$. The backward uniqueness theorem of §13 now allows us to extend this result to $\Omega\times I$ and the theorem is proved when (14.4,i) holds.

The modifications necessary when (14.4,ii) holds are indicated below. The only difference is that (14.4,ii) leads to

$$8a^2\gamma^2\int t^{-2\lambda}e^{2\lambda t}\ ||v(t)||^2\ dt \leqq Q_2\gamma^2,$$

for a constant Q, independent of γ. Instead of (14.19) we obtain an inequality of form

$$0 < \lambda\iint_{\Omega_R} e^{2\lambda t}t^{-2(\lambda+1)}v_iv_i dxdt \leqq c_1\gamma^{1-2\sigma}+c_2\gamma^2, \qquad (14.20)$$

and the proof of the theorem is completed as before.

15 A priori boundedness of the solution in the spatial variables

This § presents an interesting result of Galdi & Rionero [43] which roughly states that a solution to the Navier-Stokes equations on an unbounded region cannot, for appropriately prescribed body force and boundary data, have certain types of behaviour at infinity in the spatial variables.

To commence we introduce a weighted L^p integral, using the style of Payne [146] rather than that of [43]. Hence, let $\underset{\sim}{v}$ be a solution to P_E on $\Omega\times[0,T]$, for $T(<\infty)$ a constant, and define $\Phi_{p,R}$ by

$$\Phi_{p,R} = \int_{\Omega_R} g|\underset{\sim}{v}|^{2p}dx, \tag{15.1}$$

where p is a positive integer and g is defined by

$$g = e^{-\alpha r}, \tag{15.2}$$

for $\alpha(>0)$ to be specified.

We shall require that the velocity gradients be uniformly bounded although $\underset{\sim}{v}$ is allowed to grow like r for r large, i.e. we assume there is a constant $M(>0)$ such that

$$|v_{i,j}| \leq M, \quad |\underset{\sim}{v}| \leq Mr. \tag{15.3}$$

The next result shows that this is not possible and $|\underset{\sim}{v}|$, in fact, must be bounded.

Theorem 15.1. (Galdi & Rionero [43]). Let $\underset{\sim}{v}$ satisfying (15.3) be a classical solution to P_E on $\Omega\times[0,T]$. If $|\underset{\sim}{f}|$, $|\nabla p\#|$, $|\underset{\sim}{g}|$,

$|\underset{\sim}{h}| \leqq A$, where $p\#$ denotes the pressure term in (10.1), and if on Σ, $\left|\frac{\partial v_i}{\partial n}\right| \leqq W|\underset{\sim}{v}|$, for some constants A, W ($<\infty$), then

$$|\underset{\sim}{v}| \leqq \text{constant},$$

for each (x,t) in $\Omega\times[0,T]$.

Proof. By differentiation of $\Phi_{p,R}$,

$$\begin{aligned}\Phi'_{p,R} &= 2p\int_{\Omega_R} g|v|^{2(p-1)} v_i v_{i,t}\,dx,\\ &= 2p\int_{\Omega_R} g|v|^{2(p-1)} v_i(-v_m v_{i,m}+f_i-p\#_{,i}+\Delta v_i)\,dx,\end{aligned} \tag{15.4}$$

where in the last line $v_{i,t}$ has been substituted from (10.1).

Since $v_{i,j}$ satisfy the first of conditions (15.3), the first term on the right of (15.4) may be bounded above by

$$2pM\Phi_{p,R}. \tag{15.5}$$

The fourth term on the right may be rearranged to give

$$\begin{aligned}2p\int_{\Omega_R} g|v|^{2(p-1)} v_i\Delta v_i\,dx &= 2p\int_{\Sigma_R} gn_k v_{i,k} v_i|v|^{2(p-1)}\,dS\\ &-2p\int_{\Omega_R} g|v|^{2(p-1)} v_{i,j}v_{i,j}\,dx-4p(p-1)\int_{\Omega_R} g|v|^{2(p-2)} v_i v_{i,j} v_k v_{k,j}\,dx\\ &- 2p\int_{\Omega_R} g_{,j} v_i v_{i,j}|v|^{2(p-1)}\,dx,\end{aligned}$$

where $\Sigma_R = \Gamma_R \cup \Sigma$. The last term in this expression is then further rearranged as

$$\begin{aligned}-2p\int_{\Omega_R} g_{,j} v_i v_{i,j}|v|^{2(p-1)}\,dx &= -\int_{\Sigma_R} n_i g_{,i}|v|^{2p}\,dS\\ &+\int_{\Omega_R}\Delta g|v|^{2p}\,dx.\end{aligned}$$

Combining the last two equations gives the desired expression, namely

$$2p\int_{\Omega_R} g|v|^{2(p-1)} v_i \Delta v_i dx = 2p\int_{\Sigma_R} g n_k v_{i,k} v_i |v|^{2(p-1)} dS$$

$$-\int_{\Sigma_R} n_i g_{,i} |v|^{2p} dS + \int_{\Omega_R} \Delta g |v|^{2p} dx$$

$$-2p\int_{\Omega_R} g|v|^{2(p-1)} v_{i,j} v_{i,j} dx$$

$$-p(p-1)\int_{\Omega_R} g|v|^{2(p-2)} (|v|^2)_{,i} (|v|^2)_{,i} dx. \qquad (15.6)$$

To handle the f_i and $p\#$ terms in (15.4) we appeal to Young's inequality

$$|ab| \leq \frac{|a|^m}{m} + \frac{|b|^n}{n}, \quad \frac{1}{m} + \frac{1}{n} = 1, \quad m,n \geq 1,$$

as follows,

$$\left||v|^{2(p-1)} v_i f_i\right| \leq |v|^{2p} \frac{(2p-1)}{2p} + \frac{1}{2p}|f|^{2p}, \qquad (15.7)$$

with a similar inequality for $\nabla p\#$.

We employ (15.5)-(15.7) and the inequality for the $\nabla p\#$ term in (15.4), and discard the last two terms on the right of (15.6), to arrive at

$$\Phi'_{p,R} \leq [2p(M+2)-2]\Phi_{p,R} + \int_{\Omega_R} g|f|^{2p} dx + \int_{\Omega_R} g|\nabla p\#|^{2p} dx$$

$$+2p\int_{\Sigma_R} g n_k v_{i,k} v_i |v|^{2(p-1)} dS$$

$$-\int_{\Sigma_R} n_i g_{,i} |v|^{2p} dS + \int_{\Omega_R} \Delta g |v|^{2p} dx. \qquad (15.8)$$

By calculation,

$$g_{,i} = -\alpha g \frac{x_i}{r}, \tag{15.9}$$

and from (12.7)

$$\Delta g \leqq \alpha^2 g. \tag{15.10}$$

We use these results in (15.8), together with the fact that $\left|\frac{\partial v_i}{\partial n}\right| \leqq W|v|$ on Σ, and then allow $R \to \infty$, to obtain

$$\Phi_p' \leqq [2p(M+2)+\alpha^2-2]\Phi_p + \int_\Omega g|f|^{2p}dx + \int_\Omega g|\nabla p\#|^{2p}dx$$
$$+2pW\int_\Sigma g|h|^{2p}dS + \alpha\int_\Sigma \left(\frac{x_i n_i}{r}\right) g|h|^{2p}dS. \tag{15.11}$$

Next, let $N(p) = 2p(M+2)+\alpha^2-2$, and using an integrating factor integrate (15.11) to find, for any t in [0,T],

$$\Phi_p(t) \leqq \exp(NT).(2pB+Q+\Phi_p(0)), \tag{15.12}$$

where

$$Q = \sup_{[0,T]} \{\int_\Omega g|f|^{2p}dx + \int_\Omega g|\nabla p\#|^{2p}dx\}, \tag{15.13}$$

and

$$B = \sup_{[0,T]} \{W\int_\Sigma g|h|^{2p}dS + \frac{\alpha}{2p}\int_\Sigma [\frac{x_i n_i}{r}] g|h|^{2p}dS\}. \tag{15.14}$$

For $R(<\infty)$, $\Phi_p \geqq \exp(-\alpha R)\int_{\Omega_R}|v|^{2p}dx$, and inserting this into (15.12) yields

$$\int_{\Omega_R}|v|^{2p}dx \leqq \exp(\alpha R+NT)[\Phi_p(0)+2pB+Q]. \tag{15.15}$$

Now raise both sides to the power 1/2p, note that $(a+b+c)^{1/2p} \leqq a^{1/2p} + b^{1/2p} + c^{1/2p}$, for $a,b,c \geqq 0$, and let $p \to \infty$ to find,

$$\sup_{\Omega_R} |v| \leqq e^{(M+2)T} \cdot \lim_{p \to \infty} \{[\Phi_p(0)]^{1/2p} + B^{1/2p} + Q^{1/2p}\}, \qquad (15.16)$$

$$= \text{constant, independent of } R.$$

Since R is arbitrary, it therefore follows that $\sup|v|$ is bounded above by a constant for each (x,t) in $\Omega\times[0,T]$ and the theorem is proved.

Note. It is the presence of the nonlinear convective term $v_b v_{i,b}$ which necessitates the bound on $|v_{i,k}|$. If this method is applied to the heat equation, much weaker growth behaviour may be allowed on both the solution and its gradient to achieve the equivalent result.

So far we have applied weighted energy arguments to demonstrate how to establish a variety of results for solutions to the Navier-Stokes equations. In the next two §§ we show how the method works in other fluid dynamical problems. In both cases results different from those already established are described. In §16 we show how the weighted energy method may be adapted to prove results of continuous dependence of the solution on its initial and boundary data, while in §17 a uniqueness question in magnetohydrodynamics is studied which necessitates a new choice of weight function, due to the special nature of the problem.

16 Pointwise velocity error estimates for an elliptic fluid jet

From the three-dimensional viewpoint, the motion of a fluid jet in another fluid medium presents many difficulties. For this reason, studies of such fluid jet problems have usually sought to obtain approximate solutions to the three-dimensional equations by various means. In a series of recent papers, Green, Naghdi and their associates (see [13,14,61,136] and the references therein) have presented a novel approach to the theory of jets in three-space. Basically, these writers begin with a Cosserat one-dimensional continuum model which has structure provided by deformable vectors, called directors, and thus achieve a compact but nevertheless exact theory appropriate to jet behaviour.

We shall study the theory for a semi-infinite viscous fluid jet of elliptical cross section. In our investigation with the weighted energy method we shall see that the one-dimensional spatial character of the Green-Naghdi theory is very important; it is precisely this aspect which allows us to obtain <u>pointwise</u> error bounds for the velocity field.

Green [61] has obtained a uniqueness result for a circular viscous jet. However, he assumes boundary conditions at the nozzle and at another point in the jet and while it may be possible to measure velocities at the jet nozzle, it is likely to prove difficult to make such measurements elsewhere. In this respect the weighted energy method is naturally suited to the jet problem as we are able to dispense with boundary conditions in the jet and instead only impose behaviour on the growth of the solution at infinity.

In addition to demonstrating another application of the weighted energy method, the object of this § is also to show how it may be employed to establish continuous dependence of the solution on its boundary and initial data.

Let the jet axis coincide with the z-axis of a rectangular Cartesian system. Let ϕ_1, ϕ_2 represent the semi-axes of the elliptical cross section in the (x,y) plane and denote by v, ζ_1, ζ_2, p (all functions of z and time) the velocity along the jet, the "director velocities" in the ϕ_1, ϕ_2 directions and the pressure in the jet, respectively. In this notation the equations for a straight incompressible, homogeneous, viscous jet of elliptical cross section with the axis of the jet parallel to the gravitational field are (see Caulk & Naghdi [13], §4; [14], (4.41)):

$$\phi_{\alpha t} + v\phi_{\alpha z} = \phi_\alpha \zeta_\alpha, \quad \alpha=1,2, \tag{16.1}$$

$$v_z + \zeta_1 + \zeta_2 = 0, \tag{16.2}$$

$$\pi\rho^*\delta(v_t+vv_z) = -p_z-\rho^*\pi\delta g\#+2\mu\pi(\delta v_z)_z, \tag{16.3}$$

$$\frac{1}{4}\pi\rho^*\delta\phi_\alpha^2(\zeta_{\alpha t}+v\zeta_{\alpha z}+\zeta_\alpha^2)+2\mu\pi\delta\zeta_\alpha = p+\frac{1}{4}\mu\pi(\delta\phi_\alpha^2\zeta_{\alpha z})_z, \quad \alpha=1,2, \tag{16.4}$$

where a subscript t, z denote partial differentiation with respect to t and z, and where $\delta = \phi_1\phi_2$, ρ^*, μ are positive constants and g# is the gravitational constant. In these equations (16.2) represents the constraint of incompressibility whereas (16.1,3,4) are respectively, those of continuity, momentum and director momentum. Observe in particular, that where a subscript Greek letter is repeated no summation is implied.

For the semi-infinite jet equations (16.1)-(16.4) are defined on $(1,\infty)\times(0,T]$, where $(1,\infty)$ is the unbounded spatial region occupied by the jet and $T(<\infty)$ is a constant.

To make the above equations determinate we add the boundary conditions

$$v = V(t), \quad \zeta_\alpha = Z_\alpha(t), \quad \alpha=1,2, \quad \text{at } z = 1, \tag{16.5}$$

and the initial conditions

$$\begin{aligned} &\phi_\alpha(z,0)=f_\alpha(z), \ \alpha=1,2; \quad v(z,0)=\ell(z); \\ &\zeta_1(z,0)=m(z); \quad \zeta_2(z,0)=-m(z)-\ell_z(z); \end{aligned} \tag{16.6}$$

where V, Z_α, f_α, ℓ, m are preassigned.

The boundary-initial value problem (16.1)-(16.6) is denoted by P_J.

We wish to study the dependence of the solution to P_J on its boundary and initial data and so, in accordance with standard practice, we consider also another solution to P_J, denoted by a superscript *, together with its own boundary and initial data, in general different from that for the unstarred (base) solution. It is convenient to regard the starred quantities as perturbed solutions; for example, the unstarred solutions may represent the true solution to the problem and the starred ones measured values. By obtaining pointwise estimates for the difference of these quantities, we can assert how accurately the measured values reflect the correct solution along the jet, as time progresses.

Let now $(v,\phi_\alpha,\zeta_\alpha,p)$, $(v^*,\phi^*_\alpha,\zeta^*_\alpha,p^*)$ be two solutions to P_J for possibly different boundary and initial data and define u, ω_α, ψ_α, $\tilde{p}$, $\tilde{\delta}$ by

$$u=v^*-v, \quad \omega_\alpha=\zeta^*_\alpha-\zeta_\alpha, \quad \psi_\alpha=\phi^*_\alpha-\phi_\alpha,$$
$$\tilde{p}=p^*-p, \quad \tilde{\delta}=\delta^*-\delta, \tag{16.7}$$

where, since $\delta = \phi_1\phi_2$,

$$\tilde{\delta} = \phi^*_1\psi_2 + \psi_1\phi_2. \tag{16.8}$$

From (16.1)-(16.4) the equations governing the behaviour of u, ω_α, ψ_α, $\tilde{p}$ are obtained as

$$\psi_{\alpha t}+v^*\psi_{\alpha z}+u\phi_{\alpha z} = \phi^*_\alpha\omega_\alpha+\psi_\alpha\zeta_\alpha, \quad \alpha=1,2, \tag{16.9}$$

$$u_z + \omega_1 + \omega_2 = 0, \tag{16.10}$$

$$\rho^*\pi[\delta^*(u_t+v^*u_z)+\tilde{\delta}(v_t+vv_z)+\delta^*uv_z] = -\tilde{p}_z$$
$$-\rho^*\pi g\#\tilde{\delta}+2\mu\pi[(\delta^*u_z)_z+(\tilde{\delta}v_z)_z], \tag{16.11}$$

$$\tfrac{1}{4}\pi\rho^*[\delta^*\phi^{*2}_\alpha(\omega_{\alpha t}+v^*\omega_{\alpha z}+\zeta^*_\alpha\omega_\alpha)]+\tfrac{1}{4}\pi\rho^*[\zeta_\alpha\zeta^*_\alpha\phi^{*2}_\alpha\tilde{\delta}+\zeta_\alpha\delta\phi^{*2}_\alpha\omega_\alpha$$
$$+\delta\zeta^2_\alpha(\phi_\alpha+\phi^*_\alpha)\psi_\alpha+\zeta_{\alpha t}\phi^{*2}_\alpha\tilde{\delta}+\zeta_{\alpha t}\delta(\phi_\alpha+\phi^*_\alpha)\psi_\alpha+\zeta_{\alpha z}\delta^*\phi^{*2}_\alpha u$$
$$+v\zeta_{\alpha z}\phi^{*2}_\alpha\tilde{\delta}+v\zeta_{\alpha z}\delta(\phi_\alpha+\phi^*_\alpha)\psi_\alpha] = \tilde{p}-2\pi\mu(\delta^*\omega_\alpha+\zeta_\alpha\tilde{\delta})$$
$$+\tfrac{1}{4}\mu\pi[(\phi^{*2}_\alpha\delta^*\omega_{\alpha z})_z+(\delta^*[\phi_\alpha+\phi^*_\alpha]\psi_\alpha\zeta_{\alpha z})_z+(\phi^2_\alpha\tilde{\delta}\zeta_{\alpha z})_z], \quad \alpha=1,2. \tag{16.12}$$

The boundary conditions (16.6) apply at the orifice of the jet, z = 1. In addition, it is necessary to impose restrictions on the growth behaviour of solutions at infinity. For a prescribed constant $\varepsilon(>0)$ we assume

$$\begin{aligned} &|v|,\ |v_z|,\ |v_t|,\ |\phi_{\alpha z}|,\ |\zeta_1| \text{ (and hence } |\zeta_2|), \\ &|\zeta_\alpha^*|,\ |\phi_\alpha|,\ |\phi_\alpha^*|,\ |\zeta_{\alpha t}|,\ |\zeta_{\alpha z}| \leqq \varepsilon\ , \end{aligned} \tag{16.13}$$

for all z,t in $[1,\infty)\times[0,T]$.

Although the technique we employ will allow $|v_z^*|$ to grow like z^k for any $k(>0)$, conditions (16.13) on $|\zeta_\alpha^*|$, $|v|$, taken together with equation (16.10) restrict $|v_z^*|$ to be bounded. However, v^* may grow linearly, i.e.

$$|v^*| \leqq kz, \quad \text{for a constant } k(>0). \tag{16.14}$$

It is also necessary to impose a lower bound on the perturbation quantities ϕ_α^*, and we suppose

$$\phi_\alpha^*,\ \delta^* \geqq \delta_o, \quad \text{for some } \delta_o(>0). \tag{16.15}$$

Finally, the <u>pressure difference</u> $\tilde{p}$ is allowed to <u>grow</u> in the following manner,

$$|\tilde{p}| \leqq kz^{\frac{1}{2}-\theta}, \tag{16.16}$$

where θ is any constant such that $0 < \theta < \frac{1}{2}$.

It is worth contrasting condition (16.16) with that for the three-dimensional Navier-Stokes equations, see (12.5), where the difference pressure is required to <u>decay</u> like $r^{-\frac{1}{2}-\theta}$.

We now introduce a weighted energy $E(t)$, for each $Z \geqq 1$, by

$$E(t) = \tfrac{1}{2}\int_1^Z g\delta^*[\rho^*\pi u^2 + \sum_{\alpha=1}^{2} (\psi_\alpha^2 + \tfrac{1}{4}\rho^*\pi\phi_\alpha^{*2}\omega_\alpha^2)]dz, \tag{16.17}$$

where the weight g is here chosen as

$$g(z,t) = \exp[-\alpha(t+t_o)^{\beta}z], \tag{16.18}$$

for positive constants α, β, t_o, to be specified. The inclusion of $(t+t_o)^{\beta}$ is not necessary to obtain continuous dependence results which may instead be obtained simply with $e^{-\alpha z}$; however, this particular choice allows v^* to grow like z since it may be arranged that the v^* terms all appear in the combination $g_t + v^*g_z$ and by suitable choice of t_o, β this quantity can always be chosen non-positive (cf. Rionero & Galdi [164]).

Define

$$F(z,t) = \tfrac{1}{2}\rho^*\pi u^2+\tfrac{1}{2}\sum_{\alpha=1}^{2}(\psi_\alpha^2+\tfrac{1}{4}\rho^*\pi\phi_\alpha^{*2}\omega_\alpha^2),$$

and then

$$\frac{dE}{dt} = \int_1^Z(\delta_t^*gF + \delta^*gF_t + \delta^*Fg_t)dz,$$

$$= -gFv^*\delta^*\Big|_1^Z + \int_1^Z\delta^*Fg'dz + \int_1^Z g\delta^*F'dz,$$

where $' \equiv \partial/\partial t + v^*\partial/\partial z$ and where the equation $\delta^{*\prime}+\delta^*v_z^*=0$ has been employed.

Hence,

$$\frac{dE}{dt} = -gFv^*\delta^*\Big|_1^Z + \int_1^Z\delta^*Fg'dz + \int_1^Z g\delta^*[\rho^*\pi uu'+\sum_{\alpha=1}^{2}(\psi_\alpha\psi_\alpha'+$$

$$+\tfrac{1}{4}\rho^*\pi\zeta_\alpha^*\phi_\alpha^{*2}\omega_\alpha^2+\tfrac{1}{4}\rho^*\pi\phi_\alpha^{*2}\omega_\alpha\omega_\alpha')]dz,$$

$$= -gFv^*\delta^*\Big|_1^Z + \int_1^Z\delta^*Fg'dz + \int_1^Z g(-\rho^*\pi\tilde{\delta}(v_t+vv_z)u-\rho^*\pi\delta^*v_zu^2$$

$$-u\tilde{p}_z-\rho^*\pi g\#u\tilde{\delta}+2\mu\pi u[(\delta^*u_z)_z+(\tilde{\delta}v_z)_z])dz$$

$$+\int_1^Z g\delta^* \sum_{\alpha=1}^{2} \psi_\alpha(-u\phi_{\alpha z}+\phi_\alpha^*\omega_\alpha+\psi_\alpha\zeta_\alpha)dz+\int_1^Z g\sum_{\alpha=1}^{2}\omega_\alpha\{-\frac{1}{4}\pi\rho^*[\zeta_\alpha\zeta_\alpha^*\phi_\alpha^{*2}\tilde{\delta}$$

$$+\zeta_\alpha\phi_\alpha^{*2}\omega_\alpha\delta+\delta\zeta_\alpha^2(\phi_\alpha+\phi_\alpha^*)\psi_\alpha+\zeta_{\alpha t}\tilde{\delta}\phi_\alpha^{*2}+\zeta_{\alpha t}\psi_\alpha(\phi_\alpha+\phi_\alpha^*)\delta+u\zeta_{\alpha z}\phi_\alpha^{*2}\delta^*$$

$$+v\tilde{\delta}\zeta_{\alpha z}\phi_\alpha^{*2}+v\delta\zeta_{\alpha z}\psi_\alpha(\phi_\alpha+\phi_\alpha^*)] + \tilde{p} - 2\pi\mu[\delta^*\omega_\alpha+\zeta_\alpha\tilde{\delta}]$$

$$+\frac{1}{4}\mu\pi[(\phi_\alpha^{*2}\delta^*\omega_{\alpha z})_z+(\delta^*[\phi_\alpha+\phi_\alpha^*]\psi_\alpha\zeta_{\alpha z})_z+(\phi_\alpha^2\tilde{\delta}\zeta_{\alpha z})_z]\}dz, \qquad (16.19)$$

where to obtain the last expression, (16.9)-(16.12) have been used.

As with other fluid theories the pressure terms present special difficulties. In this situation we proceed with these terms in the following manner;

$$-\int_1^Z gu\tilde{p}_z dz+\int_1^Z g(\omega_1+\omega_2)\tilde{p}dz = -gu\tilde{p}\Big|_1^Z + \int_1^Z g_z u\tilde{p}dz$$

$$+\int_1^Z g(u_z+\omega_1+\omega_2)\tilde{p}dz, \qquad (16.20)$$

and the last term now vanishes because of (16.10).

The dissipative terms in (16.19) are important and we single these out below;

$$2\pi\mu\int_1^Z gu(\delta^*u_z)_z dz-2\mu\pi\int_1^Z g\delta^*(\omega_1^2+\omega_2^2)dz$$

$$+\frac{1}{4}\mu\pi\int_1^Z g\sum_{\alpha=1}^{2}\omega_\alpha(\phi_\alpha^{*2}\delta^*\omega_{\alpha z})_z dz$$

$$= -\mu\pi\int_1^Z g(2\delta^*u_z^2+\delta^*\sum_{\alpha=1}^{2}[2\omega_\alpha^2+\frac{1}{4}\phi_\alpha^{*2}\omega_{\alpha z}^2])dz-2\mu\pi\int_1^Z g_z u\delta^*u_z dz$$

$$-\frac{1}{4}\mu\pi\int_1^Z g_z\sum_{\alpha=1}^{2}(\phi_\alpha^{*2}\delta^*\omega_{\alpha z}\omega_\alpha)dz+2\mu\pi gu\delta^*u_z\Big|_1^Z$$

$$+\frac{1}{4}\mu\pi g\sum_{\alpha=1}^{2}\omega_\alpha(\phi_\alpha^{*2}\delta^*\omega_{\alpha z})\Big|_1^Z . \qquad (16.21)$$

The first integral on the right of (16.21) allows us to balance awkward quantities in other terms.

Equations (16.20) and (16.21) are inserted into (16.19) and then after a routine calculation we may deduce

$$\frac{dE}{dt} \leqq Q(1,t) + \int_1^\infty g_z u\tilde{p}dz + \int_1^\infty \delta^* Fg'dz + AE, \tag{16.22}$$

where E(t) is now the integral in (16.17) but over $(1,\infty)$, A is a constant dependent on the parameters in the problem (see [49]) and Q is defined below.

$$Q(1,t) = gv^*\delta^*F + gu\tilde{p} - 2\mu\pi gu(\delta^*u_z + \tilde{\delta}v_z) - \frac{1}{4}\mu\pi g \sum_{\alpha=1}^{2} \omega_\alpha(\phi_\alpha^{*2}\delta^*\omega_{\alpha z}$$
$$+\delta^*[\phi_\alpha + \phi_\alpha^*]\psi_\alpha\zeta_{\alpha z} + \phi_\alpha^2\tilde{\delta}\zeta_{\alpha z}), \tag{16.23}$$

where all the terms on the right hand side of this equation are to be evaluated at z = 1, t = t.

To deal with the third term on the right of (16.22) we note that

$$g_t + v^*g_z = -\alpha(t+t_o)^\beta zg\left(\frac{\beta}{t+t_o} + \frac{v^*}{z}\right)$$
$$\leqq z\alpha(t+t_o)^\beta[k - \beta/(t+t_o)],$$

where estimate (16.14) for v^* has been employed. Provided β is chosen so that $\beta \geqq k(t_o+T)$, this term may be discarded from (16.22) and there remains

$$\frac{dE}{dt} \leqq Q(1,t) + AE + \int_1^\infty g_z u\tilde{p}dz.$$

The pressure term is dealt with in a similar manner to §12. We write

$$\int_1^\infty g_z u\tilde{p}\,dz \leq \frac{m}{\delta_o}\int_1^\infty \tfrac{1}{2}\delta^* g u^2 dz + m\alpha^2\int_1^\infty g\tilde{p}^2 dz,$$

where $m = (t_o+T)^\beta$, and then combining the last two inequalities leads to

$$\frac{dE}{dt} \leq Q(1,t) + KE + m\alpha^2\int_1^\infty g\tilde{p}^2 dz, \qquad (16.24)$$

with $K = A+m/\delta_o$. Inequality (16.24) is the same as (12.9), encountered with the Navier-Stokes equations, except for the boundary term $Q(1,t)$.

We proceed now to obtain the velocity estimates alluded to in the title of the §. Firstly, Q is estimated. Suppose $\omega<1$ is a measure of the boundary data in the sense that

$$\begin{aligned} &|u(1,t)| < \omega^2, \quad |u_z| < 2\omega, \\ &|\psi_\alpha|, |\omega_\alpha|, |\omega_{\alpha z}| < \omega. \end{aligned} \qquad (16.25)$$

Then, from (16.23) we derive

$$|Q(1,t)| \leq b_1\omega^2,$$

where b_1 is the quantity

$$b_1 = \exp(-\alpha t_o^\beta)\left[\frac{1}{4}\sup_{[0,T]}|v^*(1,t)|(4+3\rho^*\pi)+k+9\mu\pi\right].$$

In view of (16.16), the pressure term may be bounded in similar fashion to §12 to obtain

$$m\alpha^2\int_1^\infty g\tilde{p}^2 dz \leq H\alpha^{2\theta},$$

where

$$H = k^2 m t_o^{2\beta(\theta-1)} \int_0^\infty x^{1-2\theta} e^{-x} dx.$$

Hence (16.24) yields

$$\frac{dE}{dt} \leq KE + b_1\omega^2 + H\alpha^{2\theta}.$$

Finally this inequality may be integrated to arrive at

$$E(t) \leq e^{Kt}E(0) + [(b_1/K)\omega^2 + (H/K)\alpha^{2\theta}](e^{Kt}-1). \tag{16.26}$$

It is desirable to split the derivation of velocity estimates into two cases; namely, those when the initial data are zero and variations are caused only by changes in boundary data and those for which the initial data are non-zero. In the former case the E(0) term in (16.26) is zero and this enables us to establish a better estimate than in the latter case. As the principal aim is to demonstrate the use of the weighted energy method for proving continuous dependence we describe only the former situation and refer to [49] for technical details of the latter case.

Theorem 16.1 (Galdi & Straughan [49]). When the initial data are zero and the boundary data satisfy (16.25), then the difference velocity u(z,t) satisfies the pointwise estimate

$$|u(z,t)| \leq \left(\frac{8e^m\sqrt{2}.M(t)+1}{\rho^*\pi\delta_o^2}\right)^{\frac{1}{2}}\omega, \tag{16.27}$$

for each $t\in[0,T]$ and $z \leq \omega^{-1/\theta}$, where M(t) is defined by

$$M(t) = [(b_1+H)/K](e^{Kt}-1).$$

Proof. From the definition of E(t) and the bounds (16.15), note that with $\alpha = 1/z\#$, for any $z\# \geqq 1$, $z\#$ fixed,

$$E(t) \geqq \frac{\rho^*\pi e^{-m}\delta_o}{2}\left[\int_1^{z\#} u^2 dz\right], \tag{16.28}$$

and since $2(\omega_1^2+\omega_2^2) \geqq u_z^2$,

$$E(t) \geqq \frac{\rho^*\pi e^{-m}\delta_o^3}{16}\left[\int_1^{z\#} u_z^2 dz\right]. \tag{16.29}$$

Also, select $\alpha(=1/z\#) = \omega^{1/\theta}$ and then (16.26) gives, since $E(0) = 0$,

$$E(t) \leqq M(t)\omega^2. \tag{16.30}$$

Moreover,

$$|u(Z,t)|^2 = 2\int_1^Z uu_z dz + |u(1,t)|^2.$$

Therefore, from the Cauchy-Schwarz inequality,

$$|u(z,t)| \leqq \left[2\left(\int_1^{z\#} u^2 dz\right)^{\frac{1}{2}}\left(\int_1^{z\#} u_z^2 dz\right)^{\frac{1}{2}} + |u(1,t)|^2\right]^{\frac{1}{2}}, \tag{16.31}$$

for every z such that $1 \leqq z \leqq z\#$.

Estimate (16.27) follows immediately on using (16.28)-(16.30) in (16.31).

<u>Remark</u>. For practical purposes, velocity estimates consistent with Hölder continuity are desirable and to obtain these it appears necessary to impose the pressure condition (16.16). However, for uniqueness, this may be weakened to one of the form

$$|\tilde{p}| \leqq kz^{\frac{1}{2}}(\log z)^{\sigma}, \quad 0 < \sigma < \tfrac{1}{2}.$$

The proof of this result is along lines similar to those described in the next § (see also Rionero & Galdi [163]).

17 Uniqueness in geo-astrophysical magnetohydrodynamic convection problems

Instability in the magnetohydrodynamic Bénard problem has been a subject of extensive study, see e.g. Roberts [165]. As the setting for the Bénard problem usually involves a conducting fluid contained between two infinite planes it is a natural problem for investigation via methods suitable for unbounded regions. The uniqueness question for magnetohydrodynamics on unbounded regions has been studied, e.g. Kanwal [78], Galdi & Rionero [42]; however, these studies usually assume that the magnetic field is assigned on the boundary, whereas for the Bénard problem, especially those connected with astro-geophysics, one has only certain continuity requirements across the boundary as will be seen below. This is a non-trivial point when using energy methods, for the nature of the boundary conditions completely changes the character of the dissipative terms; in this regard, see note 2 at the end of this chapter.

Suppose a conducting fluid is contained between two rigid regions which may be conductors, but in many applications are insulators. The region Ω of the fluid is infinite, but we require that it be contained between two planes, say $z = \pm d$, $0 < d < \infty$. The exterior of Ω is denoted by $\hat{\Omega}$ and the two boundaries by Σ_B with the unit normal $\underset{\sim}{n}$ on Σ_B pointing out of Ω, as in the figure below;

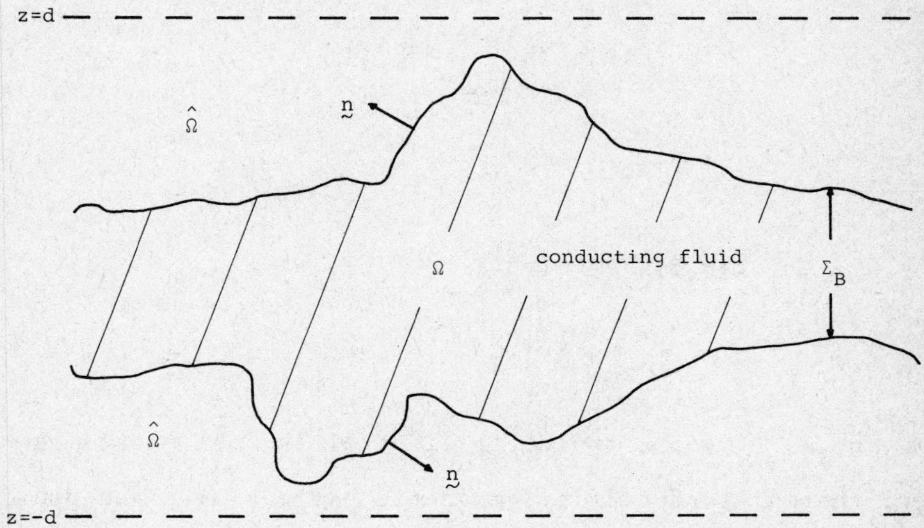

Figure 17.1

Clearly, the usual Bénard configuration is a special case of this in which Σ_B are the planes $z = \pm d$.

The equations governing a conducting fluid are given by Roberts [165], p.195 eq.(75) and p.197 eqs. (84)-(86). As we are interested in the question of uniqueness of a solution to these equations we shall consider two solutions subject to the same boundary and initial conditions. Let, therefore, $(\underset{\sim}{v},\underset{\sim}{B},T,p)$ be one solution to these equations and $(\underset{\sim}{v}^*,\underset{\sim}{B}^*,T^*,p^*)$ be another.

The equations governing the behaviour of the difference solution $\underset{\sim}{u} = \underset{\sim}{v}^*-\underset{\sim}{v}$, $\underset{\sim}{b} = \underset{\sim}{B}^*-\underset{\sim}{B}$, $\theta = T^*-T$, are found from the above equations of Roberts to be

$$u_{i,t}+v_j^*u_{i,j}+u_jv_{i,j} = -\frac{1}{\rho}\tilde{\omega}_{,i}+\frac{1}{\rho\mu}(B_j^*b_{i,j}+b_jB_{i,j})$$
$$+\nu\Delta u_i - \hat{g}\alpha\theta\delta_{i3}, \qquad (17.1)$$

$$\theta_{,t} + v_i^*\theta_{,i} + u_iT_{,i} = \kappa\Delta\theta, \qquad (17.2)$$

$$b_{i,t}+v_j^*b_{i,j}+u_jB_{i,j} = -[\mathrm{curl}(\eta\,\mathrm{curl}\,b)]_i + B_ju_{i,j}$$
$$+b_jv_{i,j} + b_ju_{i,j}, \qquad (17.3)$$

where α, ρ, κ, μ, $\hat{g}$ are the coefficient of thermal expansion, density, thermal conductivity, magnetic permeability and gravity, all of these being taken constant. In addition B_i, v_i, T are the magnetic, velocity and temperature fields, $\eta(x)$ is the (possibly non-constant) magnetic diffusivity and $\tilde{\omega}$ is a modified pressure given by

$$\tilde{\omega} = p^*-p + \frac{1}{2\mu}(|B^*|^2-|B|^2). \qquad (17.4)$$

The above equations hold throughout Ω and in $\hat{\Omega}$ we assume the temperature is known, so that in $\hat{\Omega}$ we have to solve

$$b_{i,t} = -(\mathrm{curl}(\eta\,\mathrm{curl}\,b))_i. \qquad (17.5)$$

On the boundaries Σ_B we have the conditions,

$$u_i = v_i = v_i^* = 0, \quad \theta = 0, \qquad (17.6)$$

while those for the magnetic field are (cf. Proctor [150], eq.6)

$$B_i,\ B_i^* \text{ continuous across } \Sigma_B,$$
$$\underset{\sim}{n}\times[\eta \operatorname{curl} \underset{\sim}{b}] = \underset{\sim}{0}, \tag{17.7}$$

where [...] denotes the jump in the value of the function across Σ_B.

The boundary-initial value problem consisting of (17.1)-(17.4) defined on $\Omega\times(0,\infty)$, (17.5) on $\hat{\Omega}\times(0,\infty)$ and (17.6), (17.7) on $\Sigma_B\times(0,\infty)$ together with initial conditions given on R^3 will be denoted by P_M. Since we are only interested in uniqueness, the initial conditions to be employed are

$$u_i \equiv 0,\ b_i \equiv 0,\ \theta \equiv 0 \quad \text{on} \quad R^3. \tag{17.8}$$

To apply the weighted energy method we define a region $\Omega_{R,z}$ as follows. Without loss of generality we select the origin of coordinates, 0, to be in Ω. Then, $\Omega_{R,z}$ is the cylinder with sides parallel to the z-axis, radius R $(=(x^2+y^2)^{\frac{1}{2}})$, centre 0, and length 2z, i.e. with ends at $\pm z$. The subsets of $\Omega_{R,z}$ contained in Ω and $\hat{\Omega}$ are denoted by V_R and $\hat{V}_{R,z}$, respectively. The boundaries containing and within $\Omega_{R,z}$ will be denoted by Σ_1, Σ_2, Σ, Σ_f, and Σ_B, and are indicated in the diagram below:

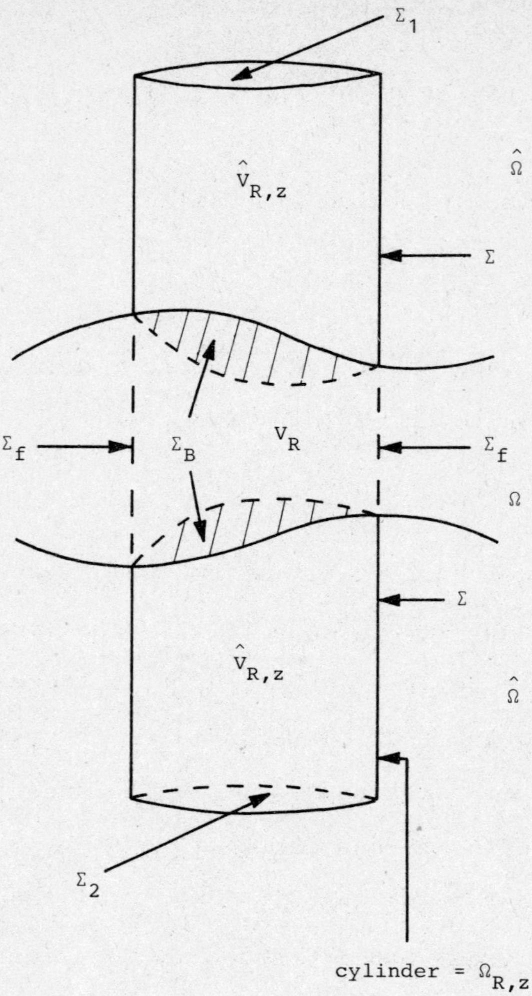

Figure 17.2

We are now in a position to state the growth conditions allowed on the solution. In most stability problems of this nature, decay conditions are imposed on the solution at infinity, see e.g. Proctor [150]. However, with the uniqueness question the weighted energy method removes the need for such restrictions and allows much weaker growth conditions. It is worth observing that in $\hat{\Omega}$ the equation for b_i is linear and so in view of the results for the heat equation described in chapter 1, we might expect that the subsequent growth behaviour allowable on b_i will be much weaker in $\hat{\Omega}$. This is, in fact, the case, although to achieve this necessitates the introduction of a different weight function to the previous §§ to take account of the various types of growth behaviour on v_i, B_i, T, p.

The solutions v_i, v_i^*, T, T*, B_i, B_i^* and $\tilde{\omega}$ are required to satisfy

$$
\begin{aligned}
&|v^*|,\ |v_{i,j}|,\ T,\ |B^*|,\ |B_{[i,j]}| \leqq D,\\
&|v|,\ |B| = O(r), \text{ as } r \to \infty,\\
&T^*,\ |\theta_{,i}|,\ |B^*_{i,j}|,\ |v^*_{i,j}| = O(r^k), \text{ any } k>0, \text{ as } r\to\infty, \qquad (17.9)\\
&|\tilde{\omega}| \leqq D, \quad r \leqq e,\\
&|\tilde{\omega}| \leqq D(\log r)^{\sigma}, \quad (0 < \sigma < \tfrac{1}{2}),\ r > e,
\end{aligned}
$$

where $D(<\infty)$ is a constant, $r^2 = x^2+y^2$; conditions (17.9) holding in $\Omega\times[0,T]$. Also,

$$|B|,\ |B^*|,\ |B_{i,j}|,\ |B^*_{i,j}| \leqq De^{\zeta|z|}, \qquad (17.10)$$

for <u>any preassigned</u> constant $\zeta(>0)$, in $\hat{\Omega}\times[0,T]$; it is understood that the growth behaviour in r, in $\hat{\Omega}$, on the quantities in (17.10) is as in (17.9). Additionally η is assumed bounded everywhere and we set

$$\bar{\eta} = \sup_{x\in R^3} \eta(x). \qquad (17.11)$$

The energy function to be employed in this section is

$$F(t) = \tfrac{1}{2}\int_{\Omega} g(\rho\mu u_i u_i + \theta^2)dx + \tfrac{1}{2}\int_{\Omega\cup\hat{\Omega}} g b_i b_i dx, \tag{17.12}$$

where g is defined in (17.13) below.

Let $V = R^2\times\{|z| < d\}$, $\hat{V} = R^3 - V$. Let C be the cylinder $\{r \leqq 1\}\times\{|z| \leqq d\}$ and let $\hat{C}$ be the union of the cylinders $\{r \leqq 1\}\times\{|z| > d\}$. The function g is then defined by

$$g(x) = \begin{cases} e^{-ar^{\gamma}}, & \text{in } V-C, \\ br^2 + c, & \text{in } V\cap C, \\ \exp[-ar^{\gamma}-\beta(|z|-d)], & \text{in } \hat{V}-\hat{C}, \\ (br^2+c).\exp[-\beta(|z|-d)], & \text{in } \hat{V}\cap\hat{C}, \end{cases} \tag{17.13}$$

where a, γ are positive constants,

$$b = -\tfrac{1}{2}a\gamma e^{-a}, \quad c = e^{-a}(1+\tfrac{1}{2}a\gamma), \tag{17.14}$$

and β is a positive constant with $\beta > 2\zeta$. The function g is continuous everywhere on R^3 and continuously differentiable everywhere except across the planes $z = \pm d$. The change to br^2+c in the region $r \leqq 1$ is needed since the exponentials have singular behaviour at r=0. Moreover, the function br^2+c is a modification of a similar one introduced by Rionero & Galdi [163].

We point out that the functions of a, γ are so chosen to enable us to deal with the pressure term, whereas the β part is to allow the exponential growth behaviour of $\underset{\sim}{B}$, indicated by (17.10), by choosing $\beta > 2\zeta$.

<u>Theorem 17.1</u> A classical solution $(u_i, b_i, \theta, \tilde{\omega})$ to the boundary-initial value problem P_M, with zero initial conditions (17.8) and with the base (unstarred) and perturbed (starred) solutions

satisfying (17.9) must necessarily be such that

$$u_i,\ b_i,\ \theta \equiv 0 \quad \text{in } R^3\times[0,\infty).$$

Proof. (cf. Theorem 12.1.)

Define a function $F_{R,z}(t)$ as F in (17.12) but with Ω, $\hat{\Omega}$ replaced by $\Omega_{R,z}$, i.e.

$$F_{R,z}(t) = \tfrac{1}{2}\int_{\Omega_{R,z}} g(\rho\mu u_i u_i + \theta^2)dx + \tfrac{1}{2}\int_{\Omega_{R,z}} g b_i b_i dx,$$

$$= \tfrac{1}{2}\int_{V_R} g(\mu\rho u_i u_i + \theta^2 + b_i b_i)dx + \tfrac{1}{2}\int_{\hat{V}_{R,z}} g b_i b_i dx. \qquad (17.15)$$

Then, differentiate $F_{R,z}$ with respect to t and subsequently substitute from (17.1)-(17.3) and (17.5) for the time derivatives to obtain

$$\begin{aligned}\dot{F}_{R,z} = &\int_{V_R} g\rho\mu(-u_i u_{i,j} v_j^* - u_j u_i v_{i,j} - \tfrac{1}{\rho}\tilde{\omega}_{,i} u_i)dx\\
&+ \int_{V_R} g(B_j^* u_i b_{i,j} + b_j u_i B_{i,j})dx + \rho\mu\nu\int_{V_R} g u_i \Delta u_i dx\\
&- \int_{V_R} g\alpha\theta w\hat{g}\rho\mu dx + \int_{V_R} g(-\theta u_i T_{,i} - v_i^*\theta_{,i}\theta + \kappa\theta\Delta\theta)dx\\
&- \int_{\Omega_{R,z}} g b_i (\mathrm{curl}(\eta\,\mathrm{curl}\,\underset{\sim}{b}))_i dx\\
&+ \int_{V_R} g b_i(-v_j^* b_{i,j} - u_j B_{i,j} + B_j^* u_{i,j} + b_j v_{i,j})dx. \qquad (17.16)\end{aligned}$$

Let us denote the terms on the right of (17.16) by I_1-I_7 and proceed to estimate these terms below.

Firstly,

$$I_6 = -\int_{\Omega_{R,z}} g b_i \varepsilon_{ijk}(\eta\varepsilon_{krs} b_{s,r})_{,j} dx$$

$$= \int_{\Omega_{R,z}} (gb_i)_{,j}\varepsilon_{ijk}\varepsilon_{krs}\eta b_{s,r}dx - \int_{\Sigma_B} n_j g[\eta\varepsilon_{krs}\varepsilon_{ijk}b_i b_{s,r}]dS$$

$$- \int_{\Gamma_{R,z}} n_j g\eta\varepsilon_{krs}\varepsilon_{ijk}b_i b_{s,r}dS$$

$$= - \int_{\Omega_{R,z}} gb_{i,j}\varepsilon_{ijk}\varepsilon_{krs}\eta b_{s,r}dx - \int_{\Omega_{R,z}} g_{,j}b_i\varepsilon_{ijk}\varepsilon_{krs}\eta b_{s,r}dx$$

$$- \int_{\Sigma_B} g[\eta \operatorname{curl}\underset{\sim}{b}.(\underset{\sim}{b}\times\underset{\sim}{n})]dS - \int_{\Gamma_{R,z}} g\eta \operatorname{curl}\underset{\sim}{b}.(\underset{\sim}{b}\times\underset{\sim}{n})dS$$

$$= - \int_{\Omega_{R,z}} g\eta|\operatorname{curl}\underset{\sim}{b}|^2dx - \int_{\Omega_{R,z}} \eta \operatorname{curl}\underset{\sim}{b}.(\underset{\sim}{b}\times\nabla g)dx$$

$$- \int_{\Gamma_{R,z}} g\eta \operatorname{curl}\underset{\sim}{b}.(\underset{\sim}{b}\times\underset{\sim}{n})dS, \tag{17.17}$$

since the integral over Σ_B vanishes by (17.7).

To deal with the terms I_2+I_7 we employ the fact that $\underset{\sim}{v}^* = \underset{\sim}{u} = \underset{\sim}{0}$ on Σ_B, as follows,

$$I_2+I_7 = \int_{V_R} g(b_j u_i\{B_{i,j}-B_{j,i}\}+B_j^*u_ib_{i,j}+B_j^*u_{i,j}b_i$$

$$-v_j^*b_ib_{i,j}+b_ib_jv_{i,j})dx$$

$$= \int_{V_R} g(2u_ib_jB_{[i,j]}+b_ib_jv_{i,j})dx$$

$$+ \int_{V_R} gB_j^*(u_ib_i)_{,j}dx - \tfrac{1}{2}\int_{V_R} gv_j^*(b_ib_i)_{,j}dx$$

$$= \int_{V_R} g(2u_ib_jB_{[i,j]}+b_ib_jv_{i,j})dx - \int_{V_R} g_{,j}B_j^*u_ib_idx$$

$$+\int_{\Sigma_f} gB_j^*n_ju_ib_idS - \tfrac{1}{2}\int_{\Sigma_f} n_jv_j^*gb^2dS + \tfrac{1}{2}\int_{V_R} v_j^*g_{,j}b^2dx. \tag{17.18}$$

The terms I_1, I_4 are left unaltered and I_3, I_5 are dealt with in a similar manner to I_2, I_6, I_7, although since θ and u_i are zero on Σ_B the calculation is simpler. The result of combining the rearrangements in (17.16) is

$$\begin{aligned}
\dot{F}_{R,z} &= \int_{V_R} \mu g_{,i}\tilde{\omega}u_i dx - \int_{\Sigma_f} \mu n_i g u_i \tilde{\omega} dS \\
&- \int_{V_R} \mu\rho g(v_j^* u_i u_{i,j} + u_j u_i v_{i,j} + \alpha\hat{g}\theta w)dx \\
&-\nu\mu\rho\int_{V_R} (g u_{i,j} u_{i,j} + g_{,j} u_i u_{i,j})dx + \int_{\Sigma_f} \nu\mu\rho n_j g u_i u_{i,j} dS \\
&- \int_{\Sigma_f} n_i g(\theta u_i + \tfrac{1}{2}\theta^2 v_i^* - \kappa\theta\theta_{,i})dS \\
&+ \int_{V_R} (g\theta_{,i} u_i T + g_{,i}\theta u_i T + \tfrac{1}{2} g_{,i} v_i^* \theta^2)dx \\
&- \int_{V_R} (\kappa g\theta_{,i}\theta_{,i} + \kappa g_{,i}\theta\theta_{,i})dx \\
&+ \int_{\Sigma_f} g(u_i b_i B_m^* n_m - \tfrac{1}{2} n_i v_i^* b^2)dS - \int_{\Gamma_{R,z}} g\eta \operatorname{curl}\underset{\sim}{b}.(\underset{\sim}{b}\times\underset{\sim}{n})dS \\
&- \int_{\Omega_{R,z}} (g\eta|\operatorname{curl}\underset{\sim}{b}|^2 + \eta \operatorname{curl}\underset{\sim}{b}.(\underset{\sim}{b}\times\nabla g))dx \\
&+ \int_{V_R} (\tfrac{1}{2} g_{,i} v_i^* b^2 - u_i b_i g_{,m} B_m^* + 2g u_i b_j B_{[i,j]} + g b_i b_j v_{i,j})dx.
\end{aligned} \tag{17.19}$$

We next allow $|z| \to \infty$ in $\Omega_{R,z}$ and then let $R \to \infty$ in (17.19) and recalling the restrictions (17.9)-(17.11) and the fact that $\beta > 2\zeta$, we arrive at an expression equivalent to (17.19) with V_R, $\Omega_{R,z}$, replaced by Ω, R^3, respectively, and without any surface integrals. The appropriate expressions for gradg are then inserted, and the following bounds noted,

$$\begin{aligned}
|\nabla g| &\leq ag \quad \text{in } \Omega, \\
|\nabla g| &\leq Mg \quad \text{in } \hat{\Omega},
\end{aligned} \tag{17.20}$$

where

$$M = \max\{a,\beta\}.$$

After routine use of inequality (AG), we may then arrive at

$$\dot{F} \leq k_1 F + \left|\mu\int_\Omega g_{,i}u_i\tilde{\omega}dx\right|, \tag{17.21}$$

where

$$k_1 = \max\{D^2(1/\nu+1/\kappa\rho\mu)+D(2+2/\rho\mu+a/\rho\mu)+\nu a^2+\hat{g}\alpha,$$
$$\kappa a^2+\hat{g}\alpha\mu\rho+2aD, 2D(2+a)+\bar{\eta}a^2/2, \tfrac{1}{2}M^2\bar{\eta}\}. \tag{17.22}$$

The treatment of the pressure term is very similar to that for the corresponding term in §13 and so the details are only sketched. Denote by Ω_2 the region $\Omega-\Omega_1$ where $\Omega_1 = \{x\in\Omega \mid r \leq 1\}$. Then

$$\left|\mu\int_\Omega g_{,i}u_i\tilde{\omega}dx\right| \leq |\mu a\gamma e^{-a}\int_{\Omega_1}|\underset{\sim}{u}|\tilde{\omega}dx|+|a\gamma\mu\int_{\Omega_2}gr^{\gamma-1}|\underset{\sim}{u}|\tilde{\omega}dx|$$
$$\leq \gamma\omega_1 + a\gamma\mu\int_{\Omega_2}gr^{\gamma-1}|\underset{\sim}{u}||\tilde{\omega}|dx,$$

where

$$\omega_1 = \mu ae^{-a}\int_{\Omega_1}|\underset{\sim}{u}||\tilde{\omega}|dx \quad (<\infty).$$

Following the steps in §13, the second term may be bounded by (see also Galdi & Rionero [42], §11)

$$a\gamma\mu\int_{\Omega_2}gr^{\gamma-1}|\underset{\sim}{u}||\tilde{\omega}|dx \leq \tfrac{1}{2}a\mu\int_{\Omega_2}g|u|^2dx+\frac{\gamma^2}{2}\int_{\Omega_2}gr^{2\gamma-2}\tilde{\omega}^2dx$$
$$\leq \frac{a}{\rho}F + 2\pi\gamma^2D^2d\left[\int_1^\infty e^{-ar^\gamma}r^{2\gamma-1}(\log r)^{2\sigma}dr\right],$$

where (17.9) has been employed, and finally

$$\leq \frac{a}{\rho}F + (2\sigma)^{\sigma} 2\pi \frac{d}{e} D^2 \gamma^{1-2\sigma},$$

where the two-dimensional unboundedness of Ω_2 has been used. Putting this expression together with the preceding one into (17.21) we find

$$\dot{F} \leq k_2 F + \gamma\omega_1 + \omega_2 \gamma^{1-2\sigma}, \tag{17.23}$$

with

$$k_2 = k_1 + a/\rho, \quad \omega_2 = 2\pi d D^2 (2\sigma)^{\sigma}/e.$$

Inequality (17.23) is integrated to obtain

$$F(t) \leq F(0).\exp(k_2 t) + (\gamma\omega_1 + \gamma^{1-2\sigma}\omega_2)\exp(k_2 t)/k_2.$$

Here $F(0) = 0$ and we bound $F(t)$ below by the integral of the same quantities over $\Omega_{R,z}$ for some $d < R, z < \infty$. From this we deduce

$$\int_{\Omega_{R,z}} (\rho\mu u_i u_i + \theta^2 + b_i b_i)dx \leq$$

$$\leq \exp[k_2 T + aR^{\gamma} + \beta(|z| - d)](\omega_1\gamma + \omega_2\gamma^{1-2\sigma})/k_2. \tag{17.24}$$

Finally, we argue that if u_i, b_i or θ is anywhere not zero in $\Omega_{R,z}$, the left side of (17.24) is positive and so we may select γ so small as to obtain a contradiction. The proof of the theorem is completed by extending $\Omega_{R,z}$ to R^3 with the aid of the arbitrariness of R, z.

Notes.

1. As the novel aspects of the work are the implications for uniqueness in the physically important magnetohydrodynamic problem we do not give other applications of the weighted

energy technique to problems such as continuous dependence in m.h.d. However, details of this are easily worked out employing similar ideas to those of §16.

2. Although it is quite straightforward to adapt the proof in §15 to prove similar a priori behaviour in many other fluid theories I have been unable to obtain an extension to the problem of this §. The cause of the trouble is the term which corresponds to $-(\mathrm{curl}(\eta\,\mathrm{curl}\,b))_i$ in (17.3). Thus, this is an instance where the magnetic dissipative term does not behave analytically like the Laplacian in other theories; that it is the aforementioned operator may be inferred from Franchi & Straughan [33] where similar results to those of §15 are derived for the equivalent non-magnetic problem to this §, when the temperature field extends to infinity, rather than the magnetic field.

18 Energy stability in bio-convection

This chapter changes the emphasis on the use of a spatial weight in unbounded domain fluid mechanics problems and studies the problem of determining the critical non-dimensional number at which instability can develop in a layer of micro-organisms in a growth medium. This topic is interesting both mathematically and physically. Mathematically it is very interesting because it leads to the idea that it may be possible to introduce a weighted energy, in a hydrodynamic stability problem, which is somehow the natural energy to use (although until now most studies have employed no weight, as mentioned in §10). As the details of the last sentence have not yet been completed I prefer to describe only the ideas as appropriate to the problem which motivated the analysis, that of bio-convection. Basically, the idea is that by introducing a spatial weight in the energy we are able to derive quite easily conditions which guarantee energy decay as $t \to \infty$ and thus guarantee nonlinear asymptotic stability. Hence, this is essentially a situation appropriate to case a) of §10 in which $g(x) \equiv 1$ is not the ideal choice. The first instance I know of in which a spatial weight was introduced in the energy to obtain sharp energy stability estimates is the paper of Joseph & Hung [75] (see also Joseph [74], volume 1) who use a weighted energy in their study of nonlinear stability of Couette flow between two concentric cylinders. However, the motive behind their analysis is different to the one which gave rise to the work of [50,51] on which this section is based.

The biological situation is that a suspension of micro-organisms is contained in a fluid layer, say the infinite layer between the planes $z = 0, \lambda$, with the gravitational field acting in the negative z direction. The organisms have a density greater than that of the containing fluid and also have a natural tendency to swim in the upward (increasing z) direction. If sufficient organisms are present, eventually the situation arises where the upper layer of the fluid is dominated by micro-organisms and these being of density greater than the fluid will tend to fall under the action of gravity and so a "convection-type" instability may develop.

The basic model for the description of the motion of the micro-organisms is developed in [16] and [108], and Childress, Levandowsky & Spiegel [16] analyse the equilibrium (zero velocity) solutions and linearized stability of these solutions.

If we let u_i and c be the velocity of the suspension and concentration of micro-organisms then the equations governing the perturbation to the equilibrium solution $u_i \equiv 0$, $c \equiv K(z)$, may be written in the following non-dimensional form (see [50,51])

$$\begin{aligned} &\sigma^{-1}(u_{i,t}+u_j u_{i,j}) = -p_{,i}+\Delta u_i-R\phi k_i, \\ &u_{i,i} = 0, \qquad\qquad (18.1) \\ &\sigma\phi_{,t}+u_i\phi_{,i} = -RK'w-\phi'+\phi''+\delta(\phi_{xx}+\phi_{yy}), \end{aligned}$$

where ' denotes $\partial/\partial z$, σ, δ are constants, ϕ is the perturbation concentration, $\phi_x = \partial\phi/\partial x$ etc., k_i is the unit vector in the z-direction, p is the pressure and R is a non-dimensional number something like the Rayleigh number in Bénard convection (it is essentially a measure of the concentration at the top surface and the layer depth). The function K is given by

$$K = c_o \exp(z-\lambda), \tag{18.2}$$

with c_o denoting the concentration at the upper surface of the layer, and equations (18.1) hold in the region between the planes $z = 0, \lambda$, λ constant. The simplest boundary conditions, which we consider here, are

$$\begin{aligned} &u_i = 0,\ \phi = 0 \quad \text{on} \quad z = 0,\lambda, \\ &u_i,\ \phi \text{ are periodic functions in } x \text{ and } y. \end{aligned} \tag{18.3}$$

The basic problem is then to construct an energy $E(u,\phi)$ and use it to find a largest value of $R = R_{crit}$, say, so that if $R < R_{crit}$ the energy $E(u,\phi) \to 0$, $t \to \infty$.

Obviously the linearized part of the time independent equations (18.1) is not symmetric (in the sense (La,b) = (Lb,a), where L is the appropriate linear operator). However, if we follow a suggestion in Childress, Levandowsky & Spiegel [16] and divide $(18.1)_3$ by $K'(>0)$, then the resulting linear operator L (written below) becomes symmetric:

$$L = \begin{pmatrix} \frac{1}{K'}\left[\delta\left(\frac{\partial^2}{\partial x^2}+\frac{\partial^2}{\partial y^2}\right)+\frac{\partial^2}{\partial z^2}-\frac{\partial}{\partial z}\right] & 0 & 0 & -R \\ 0 & \Delta & 0 & 0 \\ 0 & 0 & \Delta & 0 \\ -R & 0 & 0 & \Delta \end{pmatrix} \tag{18.4}$$

The calculation just described suggests we work with the weighted energy E(t) given by

$$E(t) = \frac{1}{2}\int_V \left(u_i u_i + \frac{\phi^2}{K'}\right)dx, \tag{18.5}$$

where V is the periodic "cell" we suppose the disturbance occupies, i.e. $0 < z < \lambda$, $\underset{\sim}{u}$, ϕ periodic in x, y.

We shall now derive a criterion on R which ensures energy stability. Let (,) and $\|.\|$ denote the inner product and norm on $L^2(V)$ and let D(u) denote the Dirichlet integral of $\underset{\sim}{u}$ over V, i.e.

$$D(u) = \int_V u_{i,j}u_{i,j}\,dx.$$

We divide $(18.1)_3$ by K' and calculate dE/dt to find

$$\frac{dE}{dt} = -2R(\phi,w) - D^{\#} - \frac{1}{2}\int_V \frac{w\phi^2}{K'}dx, \tag{18.6}$$

where w is u_3 and

$$D^{\#} = \int_V \frac{[\delta(\phi_x^2+\phi_y^2)+(\phi')^2]dx}{K'} + D(u). \tag{18.7}$$

Let $I = -2(\phi,w)$ and then from (18.6) it follows that

$$\frac{dE}{dt} \leq -D^{\#}R\left(\frac{1}{R} - \max_H \frac{I}{D^{\#}}\right) + \frac{1}{2}\left|\int_V \frac{w\phi^2 dx}{K'}\right|, \tag{18.8}$$

where

$$H = \{u_i,\phi \mid \operatorname{div}\underset{\sim}{u} = 0,\ u_i,\ \phi \text{ periodic in } x,y \text{ and } u_i = \phi = 0 \text{ on } z = 0,\lambda\} \quad .$$

Define R_E by $R_E^{-1} = \max_H \frac{I}{D^{\#}}$. Let us suppose $R < R_E$ and set $R_1 = R(1/R-1/R_E) > 0$. Inequality (18.8) may therefore be rewritten as

$$\frac{dE}{dt} + R_1 D^{\#} \leq \tfrac{1}{2}\left|\int_V \frac{w\phi^2 dx}{K'}\right| . \tag{18.9}$$

From (18.2), $c_o e^{-\lambda} \leq K' \leq c_o$ and so this together with the Cauchy-Schwarz inequality allows us to write

$$\tfrac{1}{2}\left|\int_V \frac{w\phi^2 dx}{K'}\right| \leq \frac{e^{\lambda}}{2c_o} \|w\| \; \|\phi^2\| . \tag{18.10}$$

Next, for a constant M(>0) we have the following Sobolev inequality (see Serrin [169], Ladyzhenskaya [103], Joseph [74] volume 1, Galdi & Straughan [51])

$$\|\phi^2\|^2 \leq M \|\phi\| \; \|\nabla\phi\|^3 . \tag{18.11}$$

If we use this in the $\|\phi^2\|$ term on the right of (18.10) together with the lower bound for K' we may then replace the right hand side of (18.10) by

$$\frac{e^{\lambda}M^{\frac{1}{2}}}{2c_o^{\frac{1}{2}}} \max\{1,\delta^{-1}\} \|w\| \left(\int_V \frac{\phi^2 dx}{K'}\right)^{1/4} \left(\int_V \frac{[\delta(\phi_x^2+\phi_y^2)+(\phi')^2]dx}{K'}\right)^{3/4} \tag{18.12}$$

This estimate is employed in (18.9) to derive

$$\frac{dE}{dt} + R_1 D^{\#} \leq \Lambda E^{3/4} D^{\#3/4} , \tag{18.13}$$

where

$$\Lambda = e^{\lambda} M^{\frac{1}{2}} c_o^{-\frac{1}{2}} \max\{1,\delta^{-1}\}, \tag{18.14}$$

and the Poincaré inequality is then used on part of the $R_1 D^{\#}$ term to see that

$$\frac{dE}{dt} \leq \Lambda E^{1/4} (D^{\#})^{3/4} [\frac{-R_1 \mu^{1/4}}{\Lambda} + E^{\frac{1}{2}}], \tag{18.15}$$

where μ is the Poincaré constant, i.e. $D^{\#} \geq \mu E$.

Let us now restrict attention to the class of initial data for which $\underline{E(0) < R_1^2\mu^{\frac{1}{2}}/\Lambda^2}$. Using this condition it is not difficult to see from (18.15) that $E(t) < R_1^2\mu^{\frac{1}{2}}/\Lambda^2$, for each $t > 0$, and after another application of the Poincaré inequality in (18.15), we deduce that

$$\frac{dE}{dt} \leq -R_1\mu E + \Lambda\mu^{3/4}E^{3/2}. \tag{18.16}$$

An integration of this inequality yields the bound

$$E^{\frac{1}{2}}(t) \leq \frac{1}{R_1\mu^{1/4}/\Lambda+[E^{-\frac{1}{2}}(0)-\Lambda/R_1\mu^{1/4}]\exp(R_1\mu t/2)}. \tag{18.17}$$

Under the restriction on E(0) it follows immediately that $E \to 0$, $t \to \infty$. Hence, provided the initial energy is not too large the condition $R < R_E$ is sufficient to guarantee asymptotic stability. Of course, the real problem here is to determine R_E. In [50,51] an alternative approach is adopted whereby it is shown that the maximum R_E exists and the nonlinear criterion $R < R_E$ is equivalent to the <u>linear</u> one of Childress, Levandowsky and Spiegel. Thus, the results for the linear theory show the solution is also nonlinearly stable <u>provided the restriction on E(0) is met</u>.

The above results show that if E(0) is not too great the linear and energy stability boundaries coincide (although if E(0) is large it is possible for subcritical instabilities to occur). The reason for the coincidence is that we are working with the natural (albeit weighted) energy which makes the linear time independent operator symmetric. For the ordinary Bénard problem the equivalent linear operator is symmetric at the outset and therefore the natural energy is the obvious one for which there is no weight and so no nonlinear part arises in the equivalent expression to (18.9). The upshot of this is that there is no need to restrict E(0) in that case and one rediscovers the result of Joseph [73] that the linear and energy stability boundaries in the Bénard problem are the same.

We now move away from fluid mechanics and in the final two chapters of this research note review applications of weighted energy methods to ultrahyperbolic equations and to certain models which arise in relativistic theories.

19 Solution estimates for ultrahyperbolic equations

Use of the weighted energy method in studies of ultrahyperbolic equations is very recent, having been initiated by Protter [157] and further developed by Murray & Protter [135] and Murray [133].

Let A be a second-order elliptic operator defined in a domain D in R^n and let B be another second-order elliptic operator defined in a positive cone C in R^m. Then the object is to investigate the operator

$$Lu = Au - Bu,$$

and solutions of the differential inequality

$$|Lu| \leq a|u| + b|u_{,i}|, \tag{19.1}$$

for known functions a,b. Here u is defined on $\Omega = D\times C$ and we use subscript i to denote differentiation with respect to the $x\in R^n$ variables with subscript a, b playing the same role for the $y\in R^m$ variables. The summation convention holds for repeated i or a. For boundary conditions we may choose to prescribe u on the boundary of Ω, $\partial\Omega = (D\times\partial C)\cup(\partial D\times C)$. It is revealing to examine this in a simple case and to this end we select $n = m = 2$, $A = B = I$ and $D = (0,\pi)\times(0,\pi)$, $C = (0,\infty)\times(0,\infty)$. A solution to

$$Lu = 0 \quad \text{in } \Omega,$$

$$u = 0 \quad \text{on } \partial\Omega,$$

is obviously $u \equiv 0$; however, $u = \sin nx_1 \sin nx_2 \sin ny_1 \sin ny_2$ for any integer n is another solution (this example is given in [157]). The difference from the standard hyperbolic case is that only u is given on $D\times\partial C$, i.e. $D\times\{y_1 = 0, y_2 = 0\}$, whereas in the hyperbolic situation the normal derivative is usually prescribed in addition. An interesting problem is, therefore, to determine what decay conditions for $r \to \infty$, $r^2=y_1^2+y_2^2$, need to be imposed to ensure $u\equiv 0$ is the only solution. It is precisely problems of this nature which were investigated by Protter [157]. We shall review one of his theorems, which typifies the results obtained and the method of analysis employed.

Consider the operator

$$Lu = u_{,ii} - u_{,aa}, \tag{19.2}$$

in $\Omega = D\times C$, where D is a bounded domain in R^n and $C = \{(y_1,y_2)\,|\,0 \leqq y_1 < \infty,\ 0 \leqq y_2 < \infty\}$. (In (19.2) $u_{,ii}$ denotes the Laplacian of u with respect to the x variables whereas $u_{,aa} = \frac{\partial^2 u}{\partial y_1^2} + \frac{\partial^2 u}{\partial y_2^2}$.) With $u = 0$ on $\partial\Omega$, Protter [157] studies the differential inequality (19.1).

<u>Theorem 19.1</u> (Protter [157]). Let u satisfy (19.1) in Ω with L as in (19.2) and suppose that $a = O(r^{-1})$, $b = O(r^{-1})$ as $r \to \infty$. Let u satisfy the boundary condition $u = 0$ on $\partial\Omega$ and suppose that

$$r^{\gamma}\int_{G_r} u_{,i}u_{,i}\,dS \to 0 \quad \text{as} \quad r \to \infty, \tag{19.3}$$

for all $\gamma > 0$, where G_r is the (n+1)-dimensional set $\Omega \cap \{\underset{\sim}{y}\ |\ y_1^2+y_2^2 = r^2\}$ with dS denoting the (n+1)-dimensional volume element on G_r. Then, $u \equiv 0$ in Ω.

Proof. Since D is bounded, for $r < \infty$, by Poincaré's inequality there is a constant c_1 with

$$\int_{G_r} u^2 dS \leq c_1 \int_{G_r} u_{,i} u_{,i} dS.$$

Therefore, by virtue of (19.3)

$$r^\gamma \int_{G_r} u^2 dS \to 0, \quad r \to \infty \quad \text{for every} \quad \gamma > 0. \tag{19.4}$$

Define $z = r^\gamma u$, γ fixed, then

$$Lu = r^{-\gamma} Lz + 2\gamma r^{-\gamma-2} y_a z_{,a} - \gamma^2 r^{-\gamma-2} z,$$

and then use the inequality $|A+B|^2 \geq 2AB$ and integrate the resulting inequality over Ω to arrive at

$$\int_\Omega r^{2\gamma+2} |Lu|^2 dV \geq 4\gamma \int_\Omega Lzy_a z_{,a} dV - 4\gamma^3 \int_\Omega r^{-2} y_a z z_{,a} dV, \tag{19.5}$$

where dV denotes integration in Ω. (Since (19.3) and (19.4) hold for any $\gamma > 0$ and Lu satisfies (19.1) the integrals in (19.5) are meaningful.) Next, the terms on the right of (19.5) are integrated by parts and after some manipulation and use of (19.3), (19.4) and the boundary condition it may be shown that the second term on the right of (19.5) vanishes while the first reduces to

$$2\gamma \int_\Omega z_{,i} z_{,i} dV.$$

Hence, (19.5) becomes, in terms of u

$$2\gamma \int_\Omega r^{2\gamma} u_{,i} u_{,i} dV \leq \int_\Omega r^{2\gamma+2} |Lu|^2 dV. \tag{19.6}$$

However, Lu satisfies (19.1) with a, b = $O(r^{-1})$ as $r \to \infty$ and so

$$|Lu|^2 \leqq c_2 u^2 + c_3 u_{,i} u_{,i},$$

for some constants c_2, c_3. Furthermore, by virtue of the Poincaré inequality and (19.4) the left hand side of (19.6) can be written to incorporate a u^2 term and so for some constants c_4, c_5 (19.6) leads to

$$2\gamma \int_\Omega r^{2\gamma}(u^2 + u_{,i}u_{,i})\,dV \leqq \int_\Omega r^{2\gamma}(c_4 u^2 + c_5 u_{,i}u_{,i})\,dV. \qquad (19.7)$$

A contradiction is obtained for large enough γ and so the theorem is true.

In addition to the above result Protter [157] shows how the boundary condition may be weakened and how the $\underset{\sim}{x}$-operator $\partial^2/\partial x_i \partial x_i$ may be generalized and yet still retain the validity of the theorem. He also establishes the following result for the problem studied in theorem 19.1, but with a,b only bounded.

<u>Theorem 19.2</u> (Protter [157]). Let u satisfy (19.1) in Ω with L as in (19.2) and suppose a and b are bounded in Ω. Suppose $u = 0$ on $\partial\Omega$ and

$$\exp(2r^\gamma).\int_{G_r} u_{,i}u_{,i}\,dS \to 0, \quad r \to \infty,$$

for every $\gamma > 0$. Then $u \equiv 0$ in Ω.

We complete this § by outlining the results of Murray [133] although as the proofs of her results involve many technical estimates we choose to omit them.

Murray [133] treats the general case of L stated at the beginning of the §. D is again a bounded domain in R^n, but C is the exterior of the unit ball in R^m, so with $r = y_a y_a$,

$$C(R) = \{y \in R^m \mid 1 < r < R\}.$$

Define

$$S(R) = \{y \in R^m \mid r = R\}.$$

L is the ultrahyperbolic operator $L = A - B$ defined on $D \times C$. Suppose L satisfies the following inequality in $D \times C$,

$$|Lu| \leqq \phi_0(y)|u| + \phi_1(y)|u_{,i}| + \phi_2(y)|u_{,a}|, \tag{19.8}$$

and define a weighted energy

$$E(u,R) = R^{1-m}\int_{S(R)}\int_D (|u|^2 + u_{,a}u_{,a} + a_{ij}u_{,i}u_{,j})\,dx\,dS,$$

where a_{ij} are the coefficients in the elliptic operator A, i.e. $A \equiv \frac{\partial}{\partial x_j}(a_{ij}\frac{\partial}{\partial x_i})$. Then, Murray's results may be summarized as below.

If the ϕ_α, $\alpha = 0,1,2$, are bounded in C, then

(i) u cannot have bounded support in $D \times C$, unless $u \equiv 0$;

(ii) $E(u,R)$ cannot decay faster than $\exp(-\lambda R^2)$ for all λ, unless $u \equiv 0$;

(iii) there are constants K, λ such that

$$E(u,R) \geqq K\exp(-\lambda R^2)E(u,1).$$

If $\phi_\alpha = O(r^\beta)$ for some β with $-\frac{1}{2} < \beta < \infty$, then (i) - (iii) still hold provided $\exp(-\lambda R^2)$ is replaced by $\exp(-\lambda R^{2+2\beta})$. A uniqueness theorem is deduced for the following mixed boundary value problem,

$$Lu = f(x,y,u,u_{,i},u_{,a}) \quad \text{in } D\times C(R),$$

$$u \text{ given on } \partial D\times C(R),$$

$$u \text{ and } y_a u_{,a} \text{ given on } D\times S(R).$$

It seems reasonable to assume that results of a similar nature to those of this § should be derivable when D is an exterior region in R^n by using the ideas of Murray and Protter in conjunction with the method of Galdi and Rionero described in §12.

20 Decay estimates and nonexistence results in relativistic theories

There are some very interesting results covered by the title of this § which are in the spirit of this article, e.g. Glassey & Strauss [58,59], Glassey [56], Morawetz & Strauss [130], Morawetz [127], Weder [188-191]. To complete this research note we briefly review three of the above contributions, specifically because the tools employed therein are largely weighted energy estimates.

Morawetz [127] points out that in the scattering problem for the nonlinear Klein-Gordon equation in three space dimensions,

$$u_{tt}-\Delta u+mu+P'(u) = 0, \quad m > 0, \quad P(u) \geqq 0, \tag{20.1}$$

it is necessary to know that u and its derivatives decay for large times in a fixed spatial region, provided the initial data have finite energy. By an argument which leans heavily on a weighted energy identity she establishes one such result. Namely, with the energy in a region Ω at time t defined as

$$E(u;\Omega,t) = \int_{\Omega} [\tfrac{1}{2}(u_t^2+|\nabla u|^2+mu^2) + P(u)]dx, \tag{20.2}$$

she shows that if u is continuously differentiable, has piecewise continuous second derivatives and satisfies (20.1) with given initial data such that $E(u;R^3,0) < \infty$, and P is such that

$$sP'(s) - 2P(s) \geqq aP(s), \quad a > 0,$$

then

$$\lim_{t \to \infty} \int_\Omega u^2 dx = 0 \quad \text{and} \quad \lim_{t \to \infty} E(u;\Omega,t) = 0, \tag{20.3}$$

for any finite region Ω.

Rather than describe the proof of the above result we concentrate on results of Glassey [56], which employ similar arguments to those of Morawetz, for two systems of nonlinear Schrodinger-Hartree equations. We describe only the decay results appropriate to the Cauchy problem for the equation

$$\begin{aligned} i\psi_t &= \tfrac{1}{2}\Delta\psi - v(\psi)\psi, \quad x \in R^3,\ t > 0, \\ v(\psi) &= \int_{R^3} \frac{|\psi(y,t)|^2 dy}{|x-y|}. \end{aligned} \tag{20.4}$$

Glassey's decay theorem for the above system, with ψ given at $t = 0$, is:

Theorem 20.1 (Glassey [56]). Let ψ be a solution to (20.4). Then

(i) $\int_0^\infty |\psi(0,t)|^2 dt < \infty$;

(ii) $\int_0^\infty \int_{|x| \leq R} (|\psi|^2 + |\nabla\psi|^2) dx dt < \infty$;

(iii) $\int_0^\infty \int_{R^3} \frac{v_r^2}{r} dx dt < \infty$;

(iv) $\lim_{t \to \infty} \int_{|x| \leq R} |\psi|^2 dx = 0$;

where $r = |x|$, and $R > 0$ is arbitrary.

(We should emphasize that in [56] the writer additionally considers a more complicated system than (20.4) and demonstrates nonexistence of scattering for (20.4).)

The proof of the above theorem proceeds by multiplying $(20.4)_1$ by $\zeta\bar{\psi}_r$, with $\zeta(r)$ a weight function, and after some manipulation arrives at the identity

$$-\frac{d}{dt}\,\mathrm{Im}\int \zeta\psi\bar{\psi}_r dx = -\int \zeta'|\psi_r|^2 dx - 2\pi\zeta(0)|\psi(0,t)|^2$$

$$- \int \zeta|x|^{-1}(|\nabla\psi|^2 - |\psi_r|^2)dx$$

$$+ \int |\psi|^2[\tfrac{1}{2}\Delta\zeta|x|^{-1} - \zeta'|x|^{-2} + \tfrac{1}{4}\Delta\zeta']dx$$

$$- \frac{1}{8\pi}\int \zeta'|\nabla v|^2 dx - \frac{1}{4\pi}\int (\zeta|x|^{-1} - \zeta')v_r^2 dx, \qquad (20.5)$$

where all integrals are over R^3 unless otherwise indicated and it is assumed that ψ and its derivatives decay sufficiently rapidly as $r \to \infty$.

Next multiply $(20.4)_1$ by $2\bar{\psi}$, take the imaginary part and integrate over R^3 to obtain,

$$\|\psi(t)\|^2 = \text{constant}. \qquad (20.6)$$

Since $\Gamma(x-y) = -1/4\pi|x-y|$ is the fundamental solution of Laplace's equation in three dimensions

$$v(\psi) = -4\pi\int \Gamma(x-y)|\psi(y,t)|^2 dy$$

and this leads to

$$\Delta v = -4\pi|\psi|^2. \tag{20.7}$$

Using the above relation a direct calculation shows that

$$\int v\frac{\partial}{\partial t}|\psi|^2 dx = \tfrac{1}{2}\frac{d}{dt}\int v|\psi|^2 dx.$$

Hence, by multiplying $(20.4)_1$ by $2\bar{\psi}_t$ and taking the real part we may derive

$$\int|\nabla\psi|^2 dx + \int v(\psi)|\psi|^2 dx = \text{constant}. \tag{20.8}$$

In particular, (20.6) and (20.8) show that

$$||\psi||^2 + ||\nabla\psi||^2 < \infty. \tag{20.9}$$

The weight function is then chosen as

$$\zeta(r) = -\frac{(2r+1)}{2(r+1)}, \tag{20.10}$$

for which every term on the right of (20.5) is non-negative. With this choice, parts (i) - (iii) of theorem 20.1 follow quickly from (20.5) after integrating in time over $(0,\infty)$ and using (20.9) on the left hand side. (In the proof that $\int_0^\infty\int_{|x|\leq R}|\nabla\psi|^2 dxdt < \infty$ it is first shown that $\int_0^\infty\int_{|x|\leq R}|\psi_r|^2 dxdt < \infty$ and then (20.5) is used again with $\zeta \equiv 1$ to deduce the required bound.)

Part (iv) of the theorem is established using an argument similar to one in [127] which essentially uses the fact that the integral over an interval of fixed length of a continuous $L^1(0,\infty)$ function is arbitrarily small near infinity. It is worth observing that if $\bar{\psi}$ and ψ_r are uniformly bounded on $\{|x| \leqq R\}\times(0,\infty)$ then the result may be established by letting

$F(t) = \int_{|x|\leqq R} |\psi|^2 dx$ for $R > 0$ fixed and then from (20.5), since

$\frac{1}{2}\Delta\zeta|x|^{-1}-\zeta'|x|^{-2}+\Delta\zeta'/4 = (r+4)/4r(r+1)^4 > 0$, $\int_0^\infty F ds < \infty$.

Moreover, a calculation similar to that used to derive (20.6) reveals that

$$\left|\frac{dF}{dt}\right| = \left|\int_{|x|=R} \bar{\psi}\psi_r dS\right| \leqq K, \tag{20.11}$$

for some constant K known a priori by assumption. Hence $F \in L^1(0,\infty)$ and $|F'| \leqq K$ for each $t > 0$ and, therefore, $F \to 0$, $t \to \infty$.

The last result we describe is due to Glassey & Strauss [58] whose work concerns the Yang-Mills equations and employs weighted energy arguments in a neat manner to show such results as the "energy", within any cone expanding strictly slower than a light cone, decays to zero like t^{-2} as $t \to \infty$.

For <u>vectors</u> $\underset{\sim}{A}^0$, $\underset{\sim}{A}^1$, $\underset{\sim}{A}^2$, $\underset{\sim}{A}^3$, $\underset{\sim}{E}^1$, $\underset{\sim}{E}^2$, $\underset{\sim}{E}^3$, $\underset{\sim}{H}^1$, $\underset{\sim}{H}^2$, $\underset{\sim}{H}^3$, $\underset{\sim}{E}^i$ and $\underset{\sim}{H}^i$ being connected with the electric and magnetic field strengths, the Yang-Mills equations may be written (see [58], pp. 3,4):

$$\frac{\partial H^1}{\partial t} = \frac{\partial E^2}{\partial x_3}-\frac{\partial E^3}{\partial x_2}+gE^3\times A^2+gA^3\times E^2+gA^0\times H^1, \tag{20.12}$$

$$\frac{\partial H^2}{\partial t} = \frac{\partial E^3}{\partial x_1}-\frac{\partial E^1}{\partial x_3}+gE^1\times A^3+gA^1\times E^3+gA^0\times H^2, \tag{20.13}$$

$$\frac{\partial H^3}{\partial t} = \frac{\partial E^1}{\partial x_2}-\frac{\partial E^2}{\partial x_1}+gE^2\times A^1+gA^2\times E^1+gA^0\times H^3, \tag{20.14}$$

$$\frac{\partial E^1}{\partial t} = \frac{\partial H^3}{\partial x_2} - \frac{\partial H^2}{\partial x_3} + gH^2 \times A^3 + gA^2 \times H^3 + gA^0 \times E^1, \tag{20.15}$$

$$\frac{\partial E^2}{\partial t} = \frac{\partial H^1}{\partial x_3} - \frac{\partial H^3}{\partial x_1} + gH^3 \times A^1 + gA^3 \times H^1 + gA^0 \times E^2, \tag{20.16}$$

$$\frac{\partial E^3}{\partial t} = \frac{\partial H^2}{\partial x_1} - \frac{\partial H^1}{\partial x_2} + gH^1 \times A^2 + gA^1 \times H^2 + gA^0 \times E^3, \tag{20.17}$$

$$\frac{\partial A^k}{\partial t} = E^k - \frac{\partial A^0}{\partial x_k} - gA^k \times A^0, \quad k = 1,2,3, \tag{20.18}$$

$$\begin{aligned} &\sum_k [\frac{\partial E^k}{\partial x_k} + gA^k \times E^k] = 0, \\ &\sum_k [\frac{\partial H^k}{\partial x_k} + gA^k \times H^k] = 0. \end{aligned} \tag{20.19}$$

In their study Glassey & Strauss [58] obtain many conservation laws and several results on the potentials $\underset{\sim}{A}^{\mu}$ as well as estimates on the field strengths $\underset{\sim}{E}^i$ and $\underset{\sim}{H}^i$. However, as this account focusses primarily on describing new uses of the weighted energy method we include only one such result from their paper.

A fundamental identity is the energy identity

$$\frac{\partial e}{\partial t} = p_{k,k}, \tag{20.20}$$

where the energy e is given by

$$e = \tfrac{1}{2} \sum_{k=1}^{3} (|\underset{\sim}{E}^k|^2 + |\underset{\sim}{H}^k|^2), \tag{20.21}$$

and the momenta p_j are defined by

$$p_i = \varepsilon_{ijk} \underset{\sim}{H}^j . \underset{\sim}{E}^k . \tag{20.22}$$

Equation (20.20) is derived by multiplying (20.12) - (20.17) by

$\underset{\sim}{H}^1,\ldots,\underset{\sim}{H}^3$, $\underset{\sim}{E}^1,\ldots,\underset{\sim}{E}^3$ in turn and adding the six resulting scalar equations.

Three further fundamental identities are next obtained by calculating $\partial p^j/\partial t$ and using (20.12)-(20.17) to substitute for the time derivatives of $\underset{\sim}{E}^i$ and $\underset{\sim}{H}^i$. After some manipulation and use of (20.19) it may be shown that

$$\frac{\partial p^j}{\partial t} = \frac{\partial e}{\partial x_j} - \sum_{k=1}^{3} \frac{\partial}{\partial x_k}(\underset{\sim}{E}^j.\underset{\sim}{E}^k+\underset{\sim}{H}^j.\underset{\sim}{H}^k), \tag{20.23}$$

where j = 1,2,3.

The next step is to multiply (20.20) by the weight t^2+r^2, ($r^2 = |x|^2$), and to multiply each of (20.23) by $2tx_j$ (no sum on j's) and then to add the four equations obtained to deduce that

$$\frac{\partial}{\partial t}[(r^2+t^2)e+2t\sum_{j=1}^{3} x_j p^j] = \sum_{k=1}^{3} \frac{\partial n^k}{\partial x_k}, \tag{20.24}$$

where

$$n^k = (r^2+t^2)p^k+2tx_k e-2t\sum_{j=1}^{3} x_j(\underset{\sim}{E}^j.\underset{\sim}{E}^k+\underset{\sim}{H}^j.\underset{\sim}{H}^k).$$

Define now $\underset{\sim}{\omega} = \underset{\sim}{x}/|\underset{\sim}{x}|$ and then integrate (20.24) over $R^3\times[0,t]$. Provided r^2e is integrable on R^3 and n^k decays sufficiently at infinity we obtain

$$\int_{R^3}[(r^2+t^2)e+2rt\omega_k p^k]dx = \text{constant}. \tag{20.25}$$

To bound the product $\omega_k p^k$ (summed over k) use (20.22) to write

$$\omega_k p^k = \omega_r \varepsilon_{rst}\underset{\sim}{H}^s.\underset{\sim}{E}^t,$$

where the usual summation convention is adopted.

By using inequality (AG) on each term on the right it is easily seen that

$$
\begin{aligned}
|\omega_k p^k| \leqq\ & \tfrac{1}{2}\omega_1^2(|\underset{\sim}{H}^2|^2+|\underset{\sim}{H}^3|^2+|\underset{\sim}{E}^2|^2+|\underset{\sim}{E}^3|^2) \\
& \tfrac{1}{2}\omega_2^2(|\underset{\sim}{H}^1|^2+|\underset{\sim}{H}^3|^2+|\underset{\sim}{E}^1|^2+|\underset{\sim}{E}^3|^2) \\
& \tfrac{1}{2}\omega_3^2(|\underset{\sim}{H}^1|^2+|\underset{\sim}{H}^2|^2+|\underset{\sim}{E}^1|^2+|\underset{\sim}{E}^2|^2) \\
\leqq\ & e.
\end{aligned}
$$

Therefore, from (20.25)

$$\int_{R^3} (t-r)^2 e\,dx \leqq \text{constant}. \tag{20.26}$$

Finally, restrict the above integral to a cone (four dimensional) of smaller aperture than a light cone (i.e. a set $\{|x| = k+t\}$) and (20.26) yields,

$$(\varepsilon t-R)^2 \int_{|x|<R+(1-\varepsilon)t} e\,dx \leqq \text{constant}, \tag{20.27}$$

provided $\varepsilon t > R$.

Inequality (20.27) is the required decay estimate from which Glassey & Strauss [58] deduce the t^{-2} decay result alluded to earlier.

It is of interest to note that results of a similar nature for solutions to Yang-Mills equations coupled to scalar fields are derived in [59].

References

1 Acker, A. & W.Walter. On the global existence of solutions of parabolic differential equations with a singular non-linear term. Nonlinear Analysis 2[1978], 499-505.

2 Aronszajn, N. Sur l'unicité du prolongement des solutions des équations aux dérivées partielles elliptiques. C.R. Acad. Sci., Paris, 242[1956], 723-725.

3 Ball, J.M. Remarks on blow-up and nonexistence theorems for nonlinear evolution equations. Quart. J. Math. (Oxford)(2), 28[1977], 473-486.

4 Bandle, C. Isoperimetric inequalities for a class of nonlinear parabolic equations. ZAMP 27[1976], 377-384.

5 Bardos, C. & L.Tartar. Sur l'unicité retrograde des equations paraboliques et quelques questions voisines. Arch. Rational Mech. Anal. 50[1973], 10-25.

6 Beevers, C.E. Uniqueness and stability in linear viscoelasticity. ZAMP 26[1975], 177-186.

7 Berryman, J.G. Evolution of a stable profile for a class of nonlinear diffusion equations with fixed boundaries. J. Math. Phys. 18[1977], 2108-2115.

8 Berryman, J.G. Extinction time for fast diffusion. Physics Letters 72A[1979], 107-110.

9 Berryman, J.G. & C.J.Holland. Nonlinear diffusion problem arising in plasma physics. Phys. Rev. Lett. 40[1978], 1720-1722.

10 Berryman, J.G. & C.J.Holland. Evolution of a stable profile for a class of nonlinear diffusion equations II. J. Math. Phys. 19[1978], 2476-2480.

11 Bushell, P.J. On the nonexistence of solutions to the generalized swirling flow problem. Proc. Roy. Soc. Edinburgh A 72[1974], 271-273.

12 Cannon, J.R. & G.H.Knightly. Some continuous dependence theorems for viscous fluid motions. SIAM J. Appl. Math. 18[1970], 627-640.

13 Caulk, D.A. & P.M.Naghdi. On the onset of breakup in inviscid and viscous jets. J. Appl. Mech. 46[1979], 291-297.

14 Caulk, D.A. & P.M.Naghdi. The influence of twist on the motion of straight elliptical jets. Arch. Rational Mech. Anal. 69[1979], 1-30.

15 Chang, P.H. & H.A.Levine. The quenching of solutions of semilinear hyperbolic equations. SIAM J. Math. Anal. 12[1981], 893-903.

16 Childress, S., M.Levandowsky & E.A.Spiegel. Pattern formation in a suspension of swimming micro-organisms: equations and stability theory. J. Fluid Mech. 63[1975], 591-613.

17 Choquet-Bruhat, Y. Un théorème d'instabilité pour certaines equations hyperboliques non lineaires. C. R. Acad. Sci. Paris (Ser. A) 276[1973], 281-284.

18 Cohen, P.J. & M.Lees. Asymptotic decay of solutions of differential inequalities. Pacific J. Math. 11[1961], 1235-1249.

19 Cooper, J. & W.A.Strauss. Energy boundedness and decay of waves reflecting off a moving obstacle. Indiana Univ. Math. J. 25[1976], 671-690.

20 Coppel, W.A. On a differential equation of boundary layer theory. Phil. Trans. A 253[1960], 101-136.

21 Cordes, H.O. Uber die erste randwertaufgabe bei quasilinearen differentialgleighungen zweiter ordnung in mehr als zwei variablen. Math. Annalen 131[1956], 278-312.

22 Cordes, H.O. Uber die bestimmtheit der lösungen elliptischer differentialgleichungen durch anfangsvorgaben. Nachr. Akad. Wiss. Göttingen Math. Phys. Kl.IIa [1956], 239-258.

23 Drake, J.R. & J.G.Berryman. Theory of nonlinear diffusion of plasma across the magnetic field of a toroidal multipole. Phys. Fluids. 20[1977], 851-857.

24 Dunn, J.E. & R.L.Fosdick. Thermodynamics, stability and boundedness of fluids of complexity 2 and fluids of second grade. Arch. Rational Mech. Anal. 56[1974], 191-252.

25 Duvaut, G. & J.L.Lions. Les inéquations en mécanique et en physique. Dunod, Paris, 1972.

26 Edmunds, D.E. On the uniqueness of viscous flows. Arch. Rational Mech. Anal. 14[1963], 171-176.

27 Edmunds, D.E. Asymptotic behaviour of solutions of the Navier-Stokes equations. Arch. Rational Mech. Anal. 22[1966], 15-21.

28 Fabrizio, M. Problemi di unicità per le equazioni di Navier-Stokes in domini non limitati. Arch. Rational Mech. Anal. 68[1978], 171-178.

29 Foias, C. & J.C.Saut. Limite du rapport de l'enstrophie sur l'énergie pour une solution faible des équations de Navier-Stokes. C. R. Acad. Sci. Paris (Ser.I) 293[1981], 241-244.

30 Fosdick, R.L. & K.R.Rajagopal. Thermodynamics and stability of fluids of third grade. Proc. Roy. Soc. A 369[1980], 351-377.

31 Fosdick, R.L. & B.Straughan. Catastrophic instabilities and related results in a fluid of third grade. Int. J. Nonlinear Mech. 16[1981], 191-198.

32 Franchi, F. Sull'unicità delle soluzioni delle equazioni di Boussinesq modificate in base all'equazione costitutiva di Cattaneo-Fox in un dominio illimitato. Accad. Naz. Lincei 65[1978], 275-281.

33 Franchi, F. & B.Straughan. A priori boundedness and stability in nonlinear convection problems. To be published.

34 Frehse, J. A discontinuous solution of a mildly nonlinear elliptic system. Math. Z. 134[1973], 229-230.

35 Friedman, J.L. Ergosphere instability. Commun. Math. Phys. 63[1978], 243-255.

36 Friedrichs, K.O. Symmetric hyperbolic linear differential equations. Comm. Pure Appl. Math. 7[1954], 345-392.

37 Fujita, H. On the blowing up of solutions of the Cauchy problem for $u_t = \Delta u + u^{1+\alpha}$. J. Fac. Sci. Univ. Tokyo, Sect.1, Part 2, 13[1966], 109-124.

38 Fujita, H. On some nonexistence and nonuniqueness theorems for nonlinear parabolic equations. Proc. of Amer. Math. Symp. on Nonlinear Functional Analysis, Chicago, 1968.

39 Fujita, H. On nonlinear equations $\Delta u + e^u = 0$ and $\partial v/\partial t = \Delta v + e^v$. Bull. Amer. Math. Soc., 75[1969], 132-135.

40 Galdi, G.P. Variational methods for stability of fluid motions in unbounded domains. Ricerche Mat. 27[1978], 387-404.

41 Galdi, G.P. & S.Rionero. A uniqueness theorem for hydrodynamic flows in unbounded domains. Ann. Mat. Pura Appl. 108[1976], 361-366.

42 Galdi, G.P. & S.Rionero. On magnetohydrodynamic motions in unbounded domains: stability and uniqueness. Ann. Mat. Pura Appl. 115[1977], 119-154.

43 Galdi, G.P. & S.Rionero. A priori estimates, continuous dependence and stability for solutions to the Navier-Stokes equations in exterior domains. Riv. Mat. Univ. Parma 5[1979], 533-556.

44 Galdi, G.P. & S.Rionero. Continuous dependence theorems for the Navier-Stokes equations in unbounded domains by the weight function method. Q. Jl. Mech. Appl. Math. 32[1979], 149-161.

45 Galdi, G.P. & S.Rionero. Continuous dependence theorems in linear elasticity on exterior domains. Int. J. Engng. Sci. 17[1979], 521-526.

46 Galdi, G.P. & S.Rionero. On the well posedness of the equilibrium problem for linear elasticity in unbounded regions. J. Elasticity 10[1980], 333-340.

47 Galdi, G.P. & S.Rionero. On the best conditions on the gradient of pressure for uniqueness of viscous flows in the whole space. Pacific J. Math. To be published.

48 Galdi, G.P. & S.Rionero. Local estimates and stability of viscous flows in an exterior domain. Arch. Rational Mech. Anal. To be published.

49 Galdi, G.P. & B.Straughan. Pointwise velocity error estimates for an elliptic fluid jet. ZAMP. To be published.

50 Galdi, G.P. & B.Straughan. An immediate connection between linear and nonlinear stability via an appropriate choice of measure: application to convection problems. In Proc. Conference on Waves and Stability; Catania, 1981. To be published by Italian C.N.R.

51 Galdi, G.P. & B.Straughan. Exchange of stabilities and nonlinear stability: a general approach with application to convection problems. To be published.

52 Gilbarg, D. & N.S.Trudinger. Elliptic partial differential equations of second order. Springer-Verlag. Berlin-Heidelberg-New York, 1977.

53 Gilbarg, D. & H.F.Weinberger. Asymptotic properties of steady plane solutions of the Navier-Stokes equations with bounded Dirichlet integral. Ann. Scuola Norm. Sup. Pisa 5(Ser.IV), [1978], 381-404.

54 Glassey, R.T. Blow-up theorems for nonlinear wave equations. Math. Z. 132[1973], 183-203.

55 Glassey, R.T. On the blowing up of solutions to the Cauchy problem for nonlinear Schrödinger equations. J. Math. Phys. 18[1977], 1794-1797.

56 Glassey, R.T. Asymptotic behaviour of solutions to certain nonlinear Schrödinger-Hartree equations. Commun. Math. Phys. 53[1977], 9-18.

57 Glassey, R.T. Finite time blow up for solutions of nonlinear wave equations. Math. Z. 177[1981], 323-340.

58 Glassey, R.T. & W.A.Strauss. Decay of classical Yang-Mills fields. Commun. Math. Phys. 65[1979], 1-13.

59 Glassey, R.T. & W.A.Strauss. Decay of a Yang-Mills field coupled to a scalar field. Commun. Math. Phys. 67[1979], 51-67.

60 Graffi, D. Sul teorema di unicità nella dinamica dei

fluidi. Ann. Mat. Pura Appl. 50[1960], 379-387.

61 Green, A.E. On the nonlinear behaviour of fluid jets. Int. J. Engng. Sci. 14[1976], 49-63.

62 Haviland, E.K. A note on unrestricted solutions of the differential equation $\Delta u = f(u)$. J. London Math. Soc. 26[1951], 210-214.

63 Hills, R.N. On the stability of second order fluids. J. Mécanique 13[1974], 159-165.

64 Hills, R.N. & R.J.Knops. Continuous dependence for the compressible linearly viscous fluid. Arch. Rational Mech. Anal. 51[1973], 54-59.

65 Hills, R.N. & R.J.Knops. Concavity and the evolutionary properties of a class of general materials. Proc. Roy. Soc. Edinburgh A 72[1974], 239-243.

66 Hills, R.N. & R.J.Knops. Evolutionary properties of elastodynamic solutions for non-negative and indefinite strain energies. J. Elasticity 4[1974], 77-82.

67 Hills, R.N. & R.J.Knops. Qualitative results for some general classes of material behaviour. SIAM J. Appl. Math. 30[1976], 424-451.

68 Jeffrey, A. Breakdown of the solution to a completely exceptional system of hyperbolic equations. J. Math. Anal. Appl. 45[1974], 375-381.

69 Jeffrey, A. The exceptional condition and unboundedness of solutions of hyperbolic systems of conservation type. Proc. Roy. Soc. Edinburgh A 77[1977], 1-8.

70 Jeffrey, A. Lecture on nonlinear wave propagation. In Proc. Conference on Waves and Stability; Catania, 1981. To be published by the Italian C.N.R.

71 John, F. Formation of singularities of solutions of nonlinear hyperbolic partial differential equations. Comm. Pure Appl. Math. 27[1974], 377-405.

72 John, F. Blow-up of solutions of nonlinear wave equations in three space dimensions. Manuscripta Math. 28[1979], 235-268.

73 Joseph, D.D. Nonlinear stability of the Boussinesq equations by the method of energy. Arch. Rational Mech. Anal. 22[1966], 163-184.

74 Joseph, D.D. Stability of fluid motions, vols. I & II. Springer tracts in Natural Philosophy, vols. 27 & 28. Springer-Verlag: Berlin-Heidelberg-New York: 1976.

75 Joseph, D.D. & W.Hung. Contributions to the nonlinear theory of stability of viscous flow in pipes and between rotating cylinders. Arch. Rational Mech. Anal. 44[1971], 1-22.

76 Kahane, C.S. On a system of nonlinear parabolic equations arising in chemical engineering. J. Math. Anal. Appl. 53[1976], 343-358.

77 Kalantarov, V.K. & O.A.Ladyzhenskaya. Blow up theorems for quasilinear parabolic and hyperbolic equations. Trudy Mat. Inst. Steklov 69[1977], 77-102.

78 Kanwal, R.P. Uniqueness of magnetohydrodynamic flows. Arch. Rational Mech. Anal. 4[1960], 335-340.

79 Kaplan, S. On the growth of solutions of quasilinear parabolic equations. Comm. Pure Appl. Math. 16[1963], 305-330.

80 Kato, T. Blow-up of solutions of some nonlinear hyperbolic equations. Comm. Pure Appl. Math. 33[1980], 501-505.

81 Kawarda, H. On the solutions of initial boundary value problems for $u_t = u_{xx}+1/(1-u)$. Pub. RIMS Kyoto Univ. 10[1975], 729-736.

82 Keller, J.B. Electrohydrodynamics I. The equilibrium of a charged gas in a container. J. Rational Mech. Anal. 5[1956], 715-724.

83 Keller, J.B. On solutions of $\Delta u = f(u)$. Comm. Pure Appl. Math. 10[1957], 503-510.

84 Keller, J.B. On solutions of nonlinear wave equations. Comm. Pure Appl. Math. 10[1957], 523-530.

85 Khosrovshahi, G.B. Non-existence of non-trivial solutions of a Schrödinger type system SIAM J. Math. Anal. 8[1977], 998-1013.

86 Khosrovshahi, G.B. On nonexistence of L_2 solutions of second order elliptic differential equations. Applicable Analysis 12[1981], 125-137.

87 Khosrovshahi, G.B., H.A.Levine & L.E.Payne. On the positive spectrum of Schrödinger operators with long range potentials. Trans. Amer. Math. Soc. 253[1979], 211-228.

88 Knops, R.J. Logarithmic convexity and other techniques applied to problems in Continuum Mechanics. Article in, "Proceedings of symposium on non-well posed problems and logarithmic convexity". Springer lecture notes in Mathematics, vol. 316, 1973.

89 Knops, R.J., H.A.Levine & L.E.Payne. Nonexistence, instability and growth theorems for solutions of a class of abstract nonlinear equations with applications to nonlinear elastodynamics. Arch. Rational Mech. Anal. 55[1974],52-72.

90 Knops, R.J. & L.E.Payne. On the stability of the Navier-Stokes equations backward in time. Arch. Rational Mech. Anal. 29[1968], 331-335.

91 Knops, R.J. & L.E.Payne. Growth estimates for solutions of evolutionary equations in Hilbert space with applications in elastodynamics. Arch. Rational Mech. Anal. 41[1971], 363-398.

92 Knops, R.J. & L.E.Payne. Uniqueness theorems in linear elasticity. Springer tracts in Natural Philosophy, vol.19, Springer-Verlag, Berlin, 1971.

93 Knops, R.J. & L.E.Payne. On potential wells and stability in nonlinear elasticity. Math. Proc. Camb. Phil. Soc. 84[1978], 177-190.

94 Knops, R.J. & L.E.Payne. Nonexistence of global solutions in nonlinear Cauchy elastodynamics. Arch. Rational Mech. Anal. 70[1979], 125-133.

95 Knops, R.J. & L.E.Payne. Some uniqueness and continuous dependence theorems for nonlinear elastodynamics in exterior domains. To be published.

96 Knops, R.J. & B.Straughan. Non-existence of global solutions to non-linear Cauchy problems arising in mechanics. Article in,"Trends in Applications of Pure Mathematics to Mechanics". Pitman, London, 1976.

97 Knops, R.J. & B.Straughan. Convergence of solutions of the equations of dynamic linear dipolar elasticity to the solutions of classical elastodynamics. Arch. Mech. Stos. 28[1976], 431-441.

98 Knops, R.J. & B.Straughan. Continuous dependence theorems in the theory of linear elastic materials with microstructure. Int. J. Engng. Sci. 14[1976], 555-565.

99 Knops, R.J. & B.Straughan. Decay and nonexistence for sublinearly forced systems in continuum mechanics. Article in, "Nonlinear partial differential equations and their applications. Collége de France Seminar, vol.II". Ed. H.Brezis ans J.L.Lions. Pitman, London, 1982.

100 Knops, R.J. & E.W.Wilkes. Theory of elastic stability. Article in, Handbuch der Physik, vol. VIa/3, Springer-Verlag, 1973.

101 Kolodner, I.I. & R.N.Pederson. Pointwise bounds for solutions of semilinear parabolic equations. J. Differential Equations 2[1966], 353-364.

102 Koshimae, T. On the nonexistence of global solutions for a nonlinear equation arising in the creep buckling theory. J. Math. Anal. Appl. 65[1978], 744-760.

103 Ladyzhenskaya, O.A. The mathematical theory of viscous incompressible flow. Gordon and Breach, New York, 1969.

104 Ladyzhenskaya, O.A. & V.A.Solonnikov. Determination of the solutions of Navier-Stokes stationary boundary-value problems with infinite dissipation in unbounded regions. Sov. Phys. Dokl. 24[1979], 967-969.

105 Lax, P. Development of singularities of solutions of nonlinear hyperbolic partial differential equations. J. Math. Phys. 5[1964], 611-613.

106 Lees, M. Asymptotic behaviour of solutions of parabolic differential inequalities. Can. J. Math. 14[1962], 626-631.

107 Lees, M. & M.H.Protter. Unique continuation for parabolic differential equations and differential inequalities. Duke Math. J. 28[1961], 369-382.

108 Levandowsky, M., S.Childress, E.A.Spiegel & S.H.Hutner. A mathematical model of pattern formation by swimmimg micro-organisms. J. Protozool. 22[1975], 296-306.

109 Levine, H.A. Some nonexistence and instability theorems for solutions of formally parabolic equations of the form $Pu_t = -Au+F(u)$. Arch. Rational Mech. Anal. 51[1973], 371-386.

110 Levine, H.A. Instability and nonexistence of global solutions to nonlinear wave equations of the form $Pu_{tt} = -Au+F(u)$. Trans. Amer. Math. Soc. 192[1974], 1-21.

111 Levine, H.A. Some additional remarks on the nonexistence of global solutions to nonlinear wave equations. SIAM J. Math. Anal. 5[1974], 138-146.

112 Levine, H.A. A note on a nonexistence theorem for nonlinear wave equations. SIAM J. Math. Anal. 5[1974], 644-648.

113 Levine, H.A. Nonexistence of global weak solutions to some properly and improperly posed problems of mathematical physics: the method of unbounded Fourier coefficients. Math. Annalen 214[1975], 205-220.

114 Levine, H.A. Uniqueness and growth of weak solutions to certain linear differential equations in Hilbert space. J. Differential Equations 17[1975], 73-81.

115 Levine, H.A. & J.T.Montgomery. The quenching of solutions of some nonlinear parabolic equations. SIAM J. Math. Anal. 11[1980], 842-847.

116 Levine, H.A. & L.E.Payne. Nonexistence theorems for the heat equation with nonlinear boundary conditions and for the porous medium equation backward in time. J. Differential Equations 16[1974], 319-334.

117 Levine, H.A. & L.E.Payne. Some non-existence theorems for initial-boundary value problems with nonlinear boundary constraints. Proc. Amer. Math. Soc. 46[1974], 277-284.

118 Levine, H.A. & L.E.Payne. On the nonexistence of global solutions to some abstract Cauchy problems of standard and nonstandard types. Rend. Matematica 8[1975], 413-428.

119 Levine, H.A. & L.E.Payne. Growth estimates and lower bounds for solutions in nonlinear elastodynamics with indefinite strain energy. J. Elasticity 5[1975], 273-285.

120 Levine, H.A. & L.E.Payne. Nonexistence for global weak solutions for classes of nonlinear wave and parabolic equations. J. Math. Anal. Appl. 55[1976], 329-334.

121 Levine, H.A. & L.E.Payne. On the nonexistence of entire solutions to nonlinear second order elliptic equations. SIAM J. Math. Anal.

122 Marsden, J.E. & T.J.R.Hughes. Topics in the mathematical foundations of elasticity. Article in, "Nonlinear analysis and mechanics: Heriot-Watt Symposium". Res. Notes in Math. Pitman, London, 1978.

123 Marshall, B., W.Strauss & S.Wainger. L^p-L^q estimates for the Klein-Gordon equation. J. Math. pures et appl. 59[1980], 417-440.

124 McLeod, J.B. The existence and uniqueness of a similarity solution arising from separation at a free streamline. Q. Jl. Math. (Oxford)(2), 23[1972], 63-77.

125 Morawetz, C.S. The decay of solutions of the exterior initial-boundary value problem for the wave equation. Comm. Pure Appl. Math. 14[1961], 561-568.

126 Morawetz, C.S. The limiting amplitude principle. Comm. Pure Appl. Math. 15[1962], 349-361.

127 Morawetz, C.S. Time decay for the nonlinear Klein-Gordon equation. Proc. Roy. Soc. A 306[1968], 291-296.

128 Morawetz, C.S. Decay for solutions of the exterior problem for the wave equation. Comm. Pure Appl. Math. 28[1975], 229-264.

129 Morawetz, C.S. Notes on time decay and scattering for some hyperbolic problems. SIAM Regional Conference Series in Applied Mathematics, vol. 19 [1975].

130 Morawetz, C.S. & W.A.Strauss. Decay and scattering of solutions of a nonlinear relativistic wave equation. Comm. Pure Appl. Math. 25[1972], 1-31.

131 Murray, A.C. Asymptotic behaviour of solutions of hyperbolic inequalities. Trans. Amer. Math. Soc. 157[1971], 279-296.

132 Murray, A.C. Uniqueness and continuous dependence for the equations of elastodynamics without strain energy function. Arch. Rational Mech. Anal. 47[1972], 195-204.

133 Murray, A.C. Asymptotic behaviour and uniqueness for an ultrahyperbolic equation with variable coefficients. J. Differential Equations 23[1977], 200-215.

134 Murray, A.C. & M.H.Protter. The asymptotic behaviour of solutions of second order systems of partial differential equations. J. Differential Equations 13[1973], 57-80.

135 Murray, A.C. & M.H.Protter. Asymptotic behaviour and the Cauchy problem for ultrahyperbolic operators. Indiana Univ. Math. J. 24[1974], 115-130.

136 Naghdi, P.M. On the applicability of directed fluid jets to Newtonian and non-Newtonian flows. J. Non-Newtonian Fluid Mech. 5[1979], 233-265.

137 Ogawa, H. Lower bounds for solutions of hyperbolic inequalities on expanding domains. J. Differential Equations 13[1973], 384-390.

138 Osserman, R. On the inequality $\Delta u \geqq f(u)$. Pacific J. Math. 7[1957], 1641-1647.

139 Padalino, K. Numerical estimates of finite escape times for solutions of some nonlinear parabolic equations. M.Sc. Thesis, Univ. Rhode Island, 1976.

140 Pao, C.V. On the blowing-up behaviour of solutions for a parabolic boundary value problem. Applicable Analysis 10[1980], 5-13.

141 Payne, L.E. Uniqueness criteria for steady state solutions of the Navier-Stokes equations. Atti Simp. Inter. Appl. dell'Anal. alla Fisica Mat. Cagliari-Sassari, pp. 130-153, 1964.

142 Payne, L.E. On the stability of solutions of the Navier-Stokes equations and convergence to steady state. SIAM J. Appl. Math. 15[1967], 392-405.

143 Payne, L.E. Uniqueness and continuous dependence criteria for the Navier-Stokes equations. Rocky Mt. J. Maths. 2[1971], 641-660.

144 Payne, L.E. Isoperimetric inequalities, maximum principles and their applications. Lecture notes: Univ. of Newcastle-Upon-Tyne, 1972.

145 Payne, L.E. Some general remarks on improperly posed problems for partial differential equations. Article in, "Symposium on non-well posed problems and logarithmic convexity". Springer lecture notes in Math., vol. 316, 1973.

146 Payne, L.E. Improperly posed problems in partial differential equations. SIAM regional conference series in Applied Mathematics, vol. 22, 1975.

147 Payne, L.E. & D.H.Sattinger. Saddle points and instability of nonlinear hyperbolic equations. Israel J. Math. 22[1975], 273-303.

148 Pederson, R.N. On the unique continuation theorem for certain second and fourth order elliptic equations. Comm. Pure Appl. Math. 11[1958], 67-80.

149 Piepenbrink, J. & R.Redheffer. Nonlinear partial differential inequalities. Arch. Rational Mech. Anal. 36[1970], 89-121.

150 Proctor, M.R.E. On Backus' necessary condition for dynamo action in a conducting sphere. Geophys. Astrophys. Fluid Dyn. 9[1977], 89-93.

151 Protter, M.H. Uniqueness theorems for the Tricomi problem. J. Rational Mech. Anal. 2[1953], 107-114.

152 Protter, M.H. New boundary value problems for the wave equation and equations of mixed type. J. Rational Mech. Anal. 3[1954], 435-446.

153 Protter, M.H. Unique continuation for elliptic equations. Trans. Amer. Math. Soc. 95[1960], 81-91.

154 Protter, M.H. Properties of solutions of parabolic equations and inequalities. Can. J. Math. 13[1961], 331-345.

155 Protter, M.H. Asymptotic behaviour of solutions of hyperbolic inequalities. Bull. Amer. Math. Soc. 68[1962], 523-525.

156 Protter, M.H. Asymptotic behaviour and uniqueness theorems for hyperbolic operators. Symp. Part. Diff. Equations, Novosibirsk, pp. 348-353. Acad. Sci., USSR Siberian Branch, Moscow, 1963.

157 Protter, M.H. Asymptotic decay for ultrahyperbolic operators. Article in, "Contributions to analysis", Academic Press, New York & London, 1974.

158 Rauch, J. Qualitative behaviour of dissipative wave equations on bounded domains. Arch. Rational Mech. Anal. 62[1976], 77-85.

159 Rauch, J. Singularities of solutions to semilinear wave equations. J. Math. pures et appl. 58[1979], 299-308.

160 Redheffer, R. On the inequality $\Delta u \geq f(u, |\mathrm{grad} u|)$. J. Math. Anal. Appl. 1[1960], 277-299.

161 Reuter, G.E.H. & K.Stewartson. A nonexistence theorem in magneto-fluid dynamics. Phys. Fluids 4[1961], 276-277.

162 Rionero, S. On the use of weighted norms in stability questions on exterior domains. Conf. Sem. Mat. Univ. Bari, vol. 157, 1979.

163 Rionero, S. & G.P.Galdi. On the uniqueness of viscous fluid motions. Arch. Rational Mech. Anal. 62[1976], 295-301.

164 Rionero, S. & G.P.Galdi. The weight function approach to uniqueness of viscous flows in unbounded domains. Arch. Rational Mech. Anal. 69[1979], 37-52.

165 Roberts, P.H. An introduction to magnetohydrodynamics. Longmans, London, 1967.

166 Rosenbloom, P.C. Singularities of solutions of nonlinear hyperbolic equations. Bull. Amer. Math. Soc. 60[1954], p. 343, abstract 421.

167 Serrin, J. Mathematical principles of classical fluid mechanics. Handbuch der Physik, vol. VIII/1. Springer, Berlin-Gottingen-Heidelberg, 1959.

168 Serrin, J. On the stability of viscous fluid motions. Arch. Rational Mech.Anal. 3[1959], 1-13.

169 Serrin, J. The initial value problem for the Navier-Stokes equations. Article in, "Nonlinear problems". Madison: Univ. Wisconsin Press: 1963.

170 Sperb, R.P. Growth estimates in diffusion-reaction problems. Arch. Rational Mech. Anal. 75[1981], 127-145.

171 Straughan, B. Further global nonexistence theorems for abstract nonlinear wave equations. Proc. Amer. Math. Soc. 48[1975], 381-390.

172 Straughan, B. Uniqueness and stability for the conduction-diffusion solution to the Boussinesq equations backward in time. Proc. Roy. Soc. A 347[1975], 435-446.

173 Straughan, B. Global nonexistence of solutions to Ladyzhenskaya's variants of the Navier-Stokes equations backward in time. Proc. Roy. Soc. Edinburgh A, 75[1976], 165-170.

174 Straughan, B. Growth and uniqueness theorems for an abstract nonstandard wave equation. SIAM J. Math. Anal. 7[1976], 519-528.

175 Straughan, B. Uniqueness and continuous dependence theorems for the conduction-diffusion solution to the Boussinesq equations on an exterior domain. J. Math. Anal. Appl. 57[1977], 203-233.

176 Straughan, B. Growth and instability theorems for wave equations with dissipation, with applications in contemporary continuum mechanics. J. Math. Anal. Appl. 61[1977], 303-330.

177 Straughan, B. Backward uniqueness and unique continuation for solutions to the Navier-Stokes equations on an exterior domain. J. Math. pures et appl. To be published.

178 Strauss, W. La décroissance asymptotique des solutions des équations d'onde non lineaires. C.R.Acad. Sci. Paris 256[1963], 2749-2750.

179 Strauss, W. A. Dispersal of waves vanishing on the boundary of an exterior domain. Comm. Pure Appl. Math. 28[1975], 265-278.

180 Tsutsumi, M. Some nonlinear evolutionary equations of second order. Proc. Jap. Acad. 47[1971], 450-455.

181 Tsutsumi, M. Existence and nonexistence of global solutions to nonlinear parabolic equations. Res. Inst. Math. Sci. Kyoto Univ. 8[1972], 211-229.

182 Tsutsumi, M. On solutions of semilinear differential equations in a Hilbert space. Math. Japonicae 17[1972], 173-193.

183 Tsutsumi, M. Existence and nonexistence of global solutions of the first boundary value problem for a certain quasilinear parabolic equation. Funkcial Ekvac. 17[1974], 13-24.

184 Walter, W. Uber ganze Lösungen der Differentialgleichung $\Delta u = f(u)$. Jber. Deutsch. Math. Verein., 57[1955], 94-102.

185 Walter, W. & H.Rhee. Entire solutions of $\Delta^p u = f(r,u)$. Proc. Roy. Soc. Edinburgh A 82[1979], 189-192.

186 Wang, P.K.C. Explosive ray solutions of nonlinear wave-wave interacting systems. Il Nuovo Cimento 28B[1975], 56-68.

187 Wang, P.K.C. Ray and explosive solutions of nonlinear evolutional equations in Hilbert spaces. J. Math. Phys. 16[1975], 251-256.

188 Weder, R. Absence of classical lumps. Commun. Math. Phys. 57[1977], 161-164.

189 Weder, R. Absence of classical lumps in constrained systems. Commun. Math. Phys. 63[1978], 1-8.

190 Weder, R. Existence, regularity and exponential decay of finite energy solutions to gauge fields. Preprint 198, Univ. Nac. Auto. Mexico, 1979.

191 Weder, R. Existence of monopoles in SU(3) Yang-Mills-Higgs theories. Preprint 202, Univ. Nac. Auto. Mexico, 1979.

192 Wittich, H. Ganze Losungen der Differentialgleichung $\Delta u = e^u$. Math. Z. 49[1944], 579-582.

<u>Additional References</u>. The following is a list of papers I would have referred to had I been aware of them in time.

1 Cazenave, T., J.P.Dias & M.Figueira. A remark on the decay of the solutions of some Schrödinger equations. Preprint 12, CMAF Univ. Lisboa, 1981.

2 Costa, D.G. & W.A.Strauss. Energy splitting. Quart. Appl. Math. 39[1981], 351-361.

3 Dias, J.P. & M.Figueira. Conservation laws and time decay for the solutions of some nonlinear Schrödinger-Hartree equations and systems. J. Math. Anal. Appl. 84[1981], 486-508.

4 Dias, J.P. & M.Figueira. On the decay of the solutions of some nonlinear Schrödinger equations. Preprint 1, CMAF Univ. Lisboa, 1982. To be published in the Proceedings of the "Symposium on Functional Analysis and Differential Equations" in memory of J.Sebastião e Silva; Lisbon [1982].

5 Dias, J.P. & A.Haraux. Smoothing effect and asymptotic behaviour for the solutions of a nonlinear time dependent system. Proc. Roy. Soc. Edinburgh A 87[1981], 289-303.

6 Martin, C. Finite escape time for Riccati differential equations. Systems & Control Letters 1[1981], 127-131.